CAMBRIDGE MONOGRAPHS ON PHYSICS

GENERAL EDITORS

N. FEATHER, F.R.S.
Professor of Natural Philosophy in the University of Edinburgh

D. SHOENBERG, PH.D., F.R.S.
Fellow of Gonville and Caius College, Cambridge

THE PHYSICS OF THE STRATOSPHERE

THE PHYSICS OF
THE STRATOSPHERE

BY

R. M. GOODY

Professor of Meteorology, Harvard University
Formerly Fellow of St John's College, Cambridge

CAMBRIDGE
AT THE UNIVERSITY PRESS
1958

CAMBRIDGE
UNIVERSITY PRESS

University Printing House, Cambridge CB2 8BS, United Kingdom

Published in the United States of America by Cambridge University Press, New York

Cambridge University Press is part of the University of Cambridge.

It furthers the University's mission by disseminating knowledge in the pursuit of education, learning and research at the highest international levels of excellence.

www.cambridge.org
Information on this title: www.cambridge.org/9781107696068

© Cambridge University Press 1958

First printed 1954
Reprinted 1958
First published 1958
First paperback edition 2014

A catalogue record for this publication is available from the British Library

ISBN 978-1-107-69606-8 Paperback

Cambridge University Press has no responsibility for the persistence or accuracy of URLs for external or third-party internet websites referred to in this publication, and does not guarantee that any content on such websites is, or will remain, accurate or appropriate.

GENERAL PREFACE

The Cambridge Physical Tracts, out of which this series of Monographs has developed, were planned and originally published in a period when book production was a fairly rapid process. Unfortunately, that is no longer so, and to meet the new situation a change of title and a slight change of emphasis have been decided on. The major aim of the series will still be the presentation of the results of recent research, but individual volumes will be somewhat more substantial, and more comprehensive in scope, than were the volumes of the older series. This will be true, in many cases, of new editions of the Tracts, as these are republished in the expanded series, and it will be true in most cases of the Monographs which have been written since the War or are still to be written.

The aim will be that the series as a whole shall remain representative of the entire field of pure physics, but it will occasion no surprise if, during the next few years, the subject of nuclear physics claims a large share of attention. Only in this way can justice be done to the enormous advances in this field of research over the War years.

N. F.

D. S.

AUTHOR'S PREFACE

x

for example, might well consider that my treatment of dynamical matters is most inadequate. I hope, however, that those who specialize in this subject will derive some of the interest from reading this monograph that I have derived from writing it.

I am indebted to Drs G. K. Batchelor, M. V. Wilkes and T. W. Wormell for commenting upon some sections. Mr C. D. Walshaw read the manuscript with care and found many minor errors and obscurities. Mr H. E. Goody has helped me greatly with his thorough reading of the proofs.

R. M. GOODY

Cambridge 1953

CONTENTS

AUTHOR'S PREFACE   *page* ix

CHAPTER I

Introduction

1.1. The discovery of the stratosphere, p. 1.   **1.2.** Nomenclature, p. 3. **1.3.** Scope of the monograph, p. 7.   **1.4.** Tools of research, p. 9.

CHAPTER II

Temperature

2.1. Introduction, p. 19.   **2.2.** The lower stratosphere, p. 22.   **2.3.** Propagation of sound, p. 30.   **2.4.** Meteors, p. 36.   **2.5.** Rockets, p. 41. **2.6.** Confirmatory evidence, p. 46.

CHAPTER III

Composition

3.1. General survey, p. 55.   **3.2.** Water vapour, p. 58.   **3.3.** Carbon dioxide, p. 70.   **3.4.** The rare gases, p. 72.   **3.5.** Other minor constituents, p. 77.

CHAPTER IV

Ozone

4.1. Introduction, p. 81.   **4.2.** Methods of measurement, p. 86.   **4.3.** Systematic ozone measurements, p. 112.   **4.4.** Theoretical treatment, p. 117.

CHAPTER V

Winds and Turbulence

5.1. Stratospheric winds, p. 125.   **5.2.** Turbulence in the stratosphere, p. 136.

CHAPTER VI

Radiation

6.1. Introduction, p. 142.   **6.2.** Absorption by molecular bands, p. 151.
6.3. Numerical computations, p. 164.   **6.4.** The thermal structure of the atmosphere, p. 169.

REFERENCES   *page* 176

INDEX   „ 183

AUTHOR'S PREFACE

I wrote this monograph assuming that the reader
a physicist who was interested to learn something of the
a subject outside his own field of study. I should point
fore that there is a very great difference between the a
a laboratory problem and the approach to a geophysic
that matter, astronomical) problem. In geophysics, an
larly in atmospheric physics, experiments are either im
or of limited value. Controlled conditions are esser
successful experiment, while with the atmosphere it is on
to observe and to hope that the conditions relevant to
interpretation are either known or can be guessed.
surprising that this may lead to incompatible results
apparently reliable observations, each intelligently interp
there are examples of this kind in the following pages.
laboratory worker would wisely reject such results a
worthy, the atmospheric physicist must often accept th
only results he is likely to obtain.

Every research has its own difficulties, and atmospher
is not unique in the effort, expense and even personal d
that may be involved in gathering representative obs
However, once again, it differs from most branches of
working with a system which is very far from equilib
a result it is very difficult to specify just how many obs
are desirable, and, moreover, when it comes to theoreti
pretation, equilibrium hypotheses, which are normally so
turn out to be useful for a rough superficial examination

In presenting this subject to those who have not
geophysics, I have sought to emphasize characteristic
where they appeared to be instructive, but to avoid issu
seemed to me to be particularly confusing. If this monog
been written for the geophysicist, I would have had to
in greater detail the differences of opinion which exi
nomenclature, the relative importance of different top
relative reliability of different observations, etc. The metec

CHAPTER I

INTRODUCTION

1.1. The discovery of the stratosphere

Towards the end of the last century, meteorologists were actively investigating the structure of the atmosphere away from the earth's surface by using kites and balloons carrying thermometers and pressure gauges. It was normally found that the temperature decreased with height at a rate of 5–10° K./km. These results fit in well with the idea of an atmosphere well stirred by winds, since violent mixing in a dry atmosphere should lead to a lapse rate† of 10° K./km. (the dry adiabatic lapse rate). Although the average observed lapse rate is less than the dry adiabatic, the condensation of water vapour would modify the adiabatic lapse rate in just this direction. As a result there was, at this time, a comfortable feeling among meteorologists that there was nothing more to be discovered. Crossley (1934) has attempted to analyse this attitude and has given three reasons why, in those days, it was confidently expected that the temperature of the atmosphere would always decrease with height:

'(a) As the temperature of the base of the atmosphere is that of the surface of the earth, and as the outer limit of the atmosphere must approach the absolute zero of temperature, the air temperature must, on the whole, decrease from the surface outwards.

'(b) The temperature is observed to decrease with height up to 10 km. or more according to latitude.

'(c) There is also a feeling that the temperature should decrease with height because the pressure does.'

Needless to say, these arguments do not bear close examination, but nevertheless the ideas behind them must have had general acceptance.

Despite this apparently satisfactory position the French investigator Teisserenc de Bort started in 1898 to use balloons in an

---

† According to normal meteorological terminology a 'lapse' is a negative vertical temperature gradient, i.e. temperature decreasing with height. Similarly, an 'inversion' is a positive temperature gradient.

attempt to obtain reliable temperature measurements up to 14 km. Like other workers at this time, he was greatly concerned with the errors of observation, which seemed to fall under three main headings:

(a) Errors of pressure, and therefore height, measurement.

(b) Lag of the thermometers.

(c) Radiation errors, caused by the direct absorption of solar radiation at great heights where the ventilation of the thermometers was inadequate.

It was possible to make estimates of the errors arising from (a) and (b), but (c) was indeterminate, although it was known that it could be large. When it appeared to be indicated in 1898 that above 11 km. the temperature of the atmosphere during daytime ceased to decrease with height and became constant, Teisserenc de Bort naturally concluded that he was faced with radiation errors. Experiments in 1899 showed, however, that this 'isothermal layer', as he named it, still existed at night-time, and two years later enough evidence had been collected for him to be able to announce his discovery as clearly established. By 1904 Teisserenc de Bort was able to publish the results of 581 ascents, 141 of which went to 14 km., and to distinguish some general properties of the layer, e.g. that it is low over depressions but high over anticyclones. Fig. 1 shows seasonal average curves constructed from these early results.

Thus, in 1904, the discovery of the isothermal layer had been firmly consolidated, although a number of years were to pass before it was generally accepted. The care with which Teisserenc de Bort unravelled a major discovery from the errors of a very difficult experiment, by means of careful and frequent measurements, makes this research one of the finest in the history of meteorology.

The first man to enter the stratosphere for certain, and to survive, was the Belgian Piccard who ascended to 16 km. in 1931. He might, however, have been preceded. On 5 September 1862 two Englishmen, Glaisher and Coxwell, claimed to have reached a height of 11·3 km. in an open gondola without oxygen apparatus. This may have taken them into the stratosphere, but even if it did, the aeronauts were scarcely fit to make observations, since

Glaisher was unconscious and Coxwell so badly frostbitten that he had to open the release valve with his teeth. This was only one of a long series of ascents which Glaisher and Coxwell made in their 93,000 cu.ft. capacity 'Mammoth' balloon.

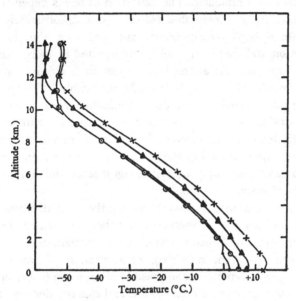

Fig. 1. The first 141 temperature records up to 14 km., grouped according to season. (After Teisserenc de Bort.) (N.B. These do not necessarily correspond exactly with the best modern measurements.) —⊙— spring; —×— summer; —△— autumn; —•— winter.

1.2. Nomenclature

It was soon discovered that the 'isothermal layer' is hardly ever exactly isothermal, and the name dropped out of general use after a decade. The most striking property of the layer was then recognized to be its great hydrostatic stability, suggesting that it might be stratified as opposed to mixed (this being the definition of the meteorological term 'stratification'). This led Teisserenc de Bort himself to suggest the word 'stratosphere' for his discovery and the word 'troposphere' for the underlying atmosphere.† Since much interest would obviously attach itself to the boundary between these regions, and since this boundary appeared to be a

† Greek *tropos* = turn; troposphere = turning or mixing sphere.

rather definite feature, it was not long before a name was found for it, the 'tropopause', suggested by Hawke and popularized by Napier Shaw.

The problem of upper-atmosphere nomenclature is now one of importance and difficulty. The definition of terms depends upon the definition of measurable features, but unfortunately most upper-atmosphere measurements are made on the limits of observation and frequently can be interpreted in more than one way. Moreover, the atmosphere is not in a steady state and measurements will never be exactly repeatable. Finally, such measurements as exist—even the voluminous records of meteorological services—are at the best only a few spot measurements in a complicated three-dimensional field. Under these circumstances care has to be exercised in the naming of a feature, since merely giving a name may imply a physical significance which later turns out to be illusory.

The difficulties of definition stem from the many different types of feature which may be measured and therefore named: thermal structure, dynamical factors (wind, etc.), concentration of major constituents (oxygen, nitrogen), concentration of minor constituents (water vapour, ozone), visible phenomena (water and ice clouds, aurorae, nightglow emissions) and electron density. There is little reason to expect detailed correspondence between these features, and fortunately there is no reason to give them all equal weight. The cornerstone of studies in atmospheric physics will always be the transformations of energy which take place, and the dynamical and thermal structures of the atmosphere are most indicative of these transformations. Dynamical factors, however, such as wind, are far too variable to be of use in general discussion, and the thermal structure is therefore the obvious choice for the purposes of nomenclature.

In fig. 2 is shown the approximate temperature distribution in the atmosphere up to 120 km. in temperate latitudes; methods of measurement will be discussed in the next chapter. Nearest to the earth's surface is the layer [a], where temperature decreases with height at about 6·5° K./km.; this is universally known as the 'troposphere'. The troposphere ends at the 'tropopause', [1], or if doubts exist as to whether this really is a well-defined feature the term 'tropopause layer' may be introduced. We will adhere to the use of the word 'tropopause', implying thereby a belief that a rather sudden change of temperature gradient at this point is the normal state of the atmosphere. The word 'stratosphere' was originally introduced to indicate the approximately isothermal region above the tropopause. We shall use the word in a wider sense, to mean that part of the atmosphere between the tropopause

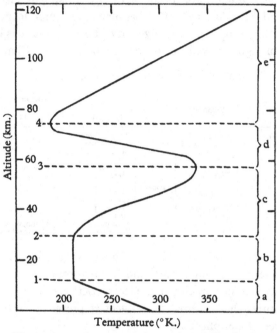

Fig. 2. Approximate vertical temperature distribution in the atmosphere in temperate latitudes.

and the inversion which starts near 75 km., [4]. By analogy with the tropopause, the feature [4] has been called the 'stratopause'. The stratosphere can conveniently be divided into the approximately isothermal region [b], or 'lower stratosphere', and the rest [c + d], or the 'upper stratosphere'. Above the stratopause lies a region whose thermal properties are not well understood, and which has mainly been investigated by means of radio techniques which have established the existence of large electron densities. It seems reasonable, therefore, that at present this region should be named after the various ionized layers which have been discovered (viz. D, E, F_1 and F_2), and that the whole region up to at least 400 km. should be called the 'ionosphere'. It has also been suggested that the upper layers of the atmosphere above 600 km., where collisions are extremely rare, should be called the 'exosphere'.

In Table I the nomenclature adopted here is compared with suggestions which have been made by Chapman (1950) and Flöhn and Penndorf (1950). Chapman gives three forms, based upon thermal properties, homogeneity (the atmosphere is believed to be thoroughly mixed below the stratopause, see Chapter III)

TABLE I. *Atmospheric nomenclature*

Reference	This book	Flöhn and Penndorf	Chapman I	Chapman II	Chapman III
a	Troposphere	Advection layer	Troposphere		
1	Tropopause	Tropopause layer	Tropopause		
b	Lower	Isothermal layer	Stratosphere		
2			Stratopause	Homosphere	Neutrosphere
c	Strato- sphere	Warm layer ⎱ Strato- sphere	Meso- incline		
3	Upper	Ozonopause	Meso- peak ⎱ Meso- sphere		
d		Upper mixing layer	Meso- decline		
4	Stratopause	Upper tropopause	Mesopause	Homopause	Neutropause
e	Ionosphere	Ionosphere	Thermosphere	Heterosphere	Ionosphere

and electrical properties. He distinguishes therefore between a 'mesopause', a 'homopause' and a 'neutropause'; but since experiment cannot distinguish clearly between the heights of these features there seems little object at present in introducing this degree of complication.

The nomenclature of Flöhn and Penndorf, although less complicated than Chapman's, is not very appropriate. The 'isothermal layer', as we have seen, is hardly ever isothermal. The 'warm layer' includes only half of the feature which might deserve this name. Finally, experiment indicates slow mixing throughout the stratosphere, and therefore the name 'upper mixing layer' conveys no useful meaning.

It is not suggested that the nomenclature adopted in this monograph has any particular physical significance.† Its main merits are that it introduces no unusual terms, so that those familiar with

† In fact Chapman's use of the word 'stratosphere' is correct physically.

the literature will recognize the meanings of terms without formal definition, and that it conveys no more than the experimental evidence justifies.

1.3. Scope of the monograph

In the following chapters we will discuss briefly the main physical properties of the stratosphere, only considering the troposphere and the ionosphere where it is essential to preserve continuity. Work upon this region of the atmosphere has mainly been concerned with measurements and calculation of the temperature, the composition, the dynamical structure and the radiation balance. With regard to temperature, we require to know not only the general vertical structure but also the variations which take place daily, seasonally, and from one latitude to another, for such information gives a valuable guide to underlying processes. Atmospheric composition is of interest from many points of view: most methods of measuring temperature depend upon a knowledge of the mean molecular mass of air; the extent to which gases separate out under gravity indicates the degree of mixing which is taking place; the radiation balance at levels below the stratopause, which creates the sources and sinks of energy necessary to maintain the atmospheric motions, is almost entirely controlled by the concentrations of three minor atmospheric constituents, viz. ozone, water vapour and carbon dioxide, which present many interesting problems of measurement; ozone is created and destroyed *in situ*, and its concentration is so intimately bound up with atmospheric conditions that the study of this gas is almost a branch of geophysics in its own right. Finally, the dynamical structure of the atmosphere is one of the clearest indications of the transformations of energy which continually take place, and the theory of radiation transfer is required to link together such studies with information upon temperatures and composition.

By limiting discussion to the stratosphere we shall not be considering many important atmospheric phenomena. As far as is known, a number of changes take place just above the stratopause. The atmosphere becomes ionized and therefore conducting, and there have been many interesting studies upon the reflexion of radio waves and short-period variations of the earth's magnetic field. Also, approximately at this level, most molecules except nitrogen begin to be dissociated by the ultra-violet radiation from the sun, leading to uncertainties in our knowledge of the mean molecular mass for air, and therefore to uncertainties in temperature measurement (see Chapter II). Further, atomic and molecular mean free paths become so great that diffusive transfer begins to predominate over mixing processes. Aurorae practically never penetrate into the stratosphere, and although some night-glow emissions are now considered to be located near 70 km., they are still generally associated with higher levels.

All these extremely interesting topics will be only briefly mentioned where they lead to information about the atmosphere near the stratopause. The reader who is interested in these problems is, however, fortunate in the recent publication of a number of books and reviews dealing with them, which are listed at the end of this monograph.

Going to the other extreme, the stratosphere is just above the levels involved in the important and complicated phenomena of weather. There is a tendency now for meteorologists to include the lower stratosphere in their dynamical considerations, although marked reactions of stratospheric conditions upon weather have yet to be demonstrated. The extremely important subject of cloud and rain formation is almost entirely restricted to the troposphere, though the highest layers of cirrus and cumulo-nimbus lie only just below the tropopause.

The remaining portion of the atmosphere, which will be described in this monograph, is neither so close that mistakes in the computation of minor perturbations are a subject for discussion in the press and on the radio, nor yet so remote that it is necessary to work with techniques akin to astronomy. It is a region whose main features can be measured and explained by fairly straight-forward physical methods, and it is the application of these methods to a field problem of great complication which will be the main concern of this book.

Finally, there will only be passing reference to the important subject of atmospheric oscillations, not because it is inappropriate, but because it has already been discussed in a monograph in this series (Wilkes, 1949).

1.4. Tools of research

Much of our knowledge of the upper atmosphere is based upon measurements made at ground level, and owing to the care and accuracy possible in the laboratory such measurements will always be of great importance. There is little in common between various ground-level experiments, except perhaps that most make use of some section of the electromagnetic spectrum, and they all need to be described individually in their context in later chapters.

In recent years it has become more and more common to attempt to make direct measurements by transporting instruments into the stratosphere on various types of vehicle. The limitations of such experiments are often imposed by the characteristics of the vehicle itself, and therefore it is worth while to discuss the important features of the three possibilities available, viz. balloons, aircraft and rockets.

1.41. Balloons

The first attempts to make measurements of the upper atmosphere employed kites and balloons. Balloons soon proved their superiority by ascending to much greater heights. The techniques of manned and unmanned balloon ascents differ very greatly, and the former have now fallen into disuse, though in the past much useful information has been gained by this method of investigation.

The main characteristic of the manned balloon ascent is the great weight that has to be carried. The observer is a considerable load, particularly since he must be protected against the low pressures of altitudes greater than 10 km., and such ascents are justifiable only if a quantity of apparatus is carried. The last great ascent was by the American helium-filled Explorer II (1938). It reached a record height of 22 km. In order to lift the weights involved, the capacity of the balloon was 3,750,000 cu.ft., although only a small fraction of this could be filled with helium at ground level since the pressure at the ceiling was only $\frac{1}{28}$ atmosphere, and a really elastic balloon is not acceptable for a manned ascent. The Explorer II had a spherical aluminium gondola in which three men could be carried. The total flight time was 8 hours 13 minutes, during which time many observations were taken, including cosmic-ray intensities at various zenith angles, the height distribution of ozone by spectroscopic methods, atmospheric ionization, the composition of the air with particular reference to oxygen, carbon dioxide and helium, the presence of spores and moulds, and air photographs for geological purposes.

Manned balloon ascents are unlikely to be of future service to research. They are difficult and costly to organize, considerable danger to human life is involved (one Russian and one American ascent have ended in disaster), and the advance of techniques has now made it possible to obtain all important results with much cheaper unmanned balloons.

The main difficulties with unmanned balloons are the light, reliable and automatic equipment required and the recovery of data after the flight. Data may be obtained either by the recovery of the whole balloon or by the radio transmission of instrument readings to the ground. The radio sonde, now universally used for routine observations, employs this second method. For research purposes the apparatus may be too complicated for this technique to be used, and the usual procedure is to attempt to recover the balloon and the attached equipment. In this country, with its dense population, a very good recovery rate, as high as 80%, may be obtained by attaching to the gondola a label with the promise of a small reward, and since the other 20% are lost partly in the sea or other inaccessible areas, the choice of favourable occasions can lead to an even better recovery rate. Alternatively, the study of upper winds and the preliminary release of a pilot balloon, together with tracking of the main balloon by telescope and radar, can give a good estimate of the point of landing, where the investigator may reasonably hope to proceed by car with a pair of binoculars and thus be present at the time of landing.

Before the last war, rubber balloons were in general use. The maximum height attained for certain with such balloons was about 30 km., but such heights were only rarely reached, 20 km. being more usual. Long experience is required to obtain good results with reasonably cheap balloons, and much of the progress has been made by a few individual investigators—for example, E. Regener, of Stuttgart, who has described some of his earlier experiences in an interesting article (Regener, 1935), to which the reader may turn for further information. Regener has sent apparatus weighing 6·5 kg. up to heights of 28 km., using four balloons, one below the other, connected with fishing-line attached to the equators of the balloons. Points of technique include: complicated launching arrangements if there is any wind at the ground, a wooden-framed silk brake which both restricts the rate of ascent at lower levels and also acts as a parachute after one of the balloons has burst, a considerable free lift to give stability, and an ingenious 'glass-house' device of cellophane above and aluminium foil below to give an almost constant temperature in the gondola at all heights. Often rubber balloons will burst at quite low altitudes on account of imperfections in the rubber. The smallest point of weakness controls the strength of the whole balloon, and minute imperfections can be detected only by examination of the stretched rubber against a strong light; imperfections found in this way can be patched, leading in general to better performance. Even under the best conditions the normal elastic limit of the rubber is never reached, since its elastic properties are impaired by the low temperatures of the stratosphere.

Many of these troubles can now be avoided by the use of neoprene-latex balloons, which are increasingly coming into use for altitudes above 20 km. They are of more uniform quality than rubber balloons and are less affected by the low temperatures. It is now a fairly straightforward matter to reach 30 km. with unmanned balloons, and the difficulties of acute weight restriction can be avoided if the investigator can afford to buy the huge balloons now being used in the U.S.A.

Fig. 3 shows an arrangement recently used to carry an ultraviolet solar spectroscope to high levels in order to measure the vertical distribution of ozone. The net free lift was 2 kg., which should have given a rate of ascent of 280 m./min.† This rate was only reached, however, near the top of the ascent; near the ground the action of the brake reduced it to 60 m./min. At 28 km. one of the balloons burst and the remaining balloon descended at almost exactly the same rate as it had ascended. The bursting diameter was 7·7 m., when the thickness of the neoprene-latex was only 9 μ.

† Simple theory indicates, and experiment confirms, that an unhampered, elastic balloon should ascend at nearly constant speed until it bursts.

Laboratory tests had indicated that the elastic limit was reached at a thickness of 5.6μ, and this performance was therefore considered to be good.

For routine work, meteorological departments in most countries have now produced radio sondes which are so cheap that their loss is not of great importance, and so light that small 3 ft. balloons are

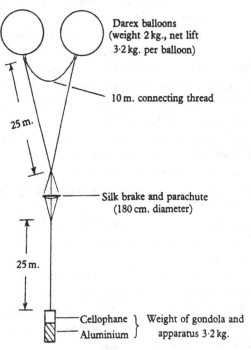

Darex balloons
(weight 2 kg., net lift
3·2 kg. per balloon)

10 m. connecting thread

25 m.

Silk brake and parachute
(180 cm. diameter)

25 m.

Cellophane ⎱ Weight of gondola and
Aluminium ⎰ apparatus 3·2 kg.

Fig. 3. A typical high-altitude balloon arrangement (schematic). (After E. Regener.) The thread has a breaking strain of 5 kg. and the total weight of thread and brake is 1·19 kg.

large enough to carry them to 20 km. The British radio sonde is fairly typical. It weighs about 3 lb. and contains elements which will react to temperature, pressure and relative humidity. Each of these elements varies the inductance of a tuned oscillator, thus generating audio-frequency signals which vary with the deflexion of the element. The three elements in turn are switched into a circuit where the audio-frequency signal modulates a 27 Mc./sec. carrier, which is transmitted. The humidity measurements are probably of no significance above 7 km., but the temperature measurements are as accurate as is required for routine work, provided that sufficient trouble is taken to avoid radiation errors.

1.42. *Aircraft*

Piston-engined aircraft can be made to climb as high as 16 km., and presumably jet-engined aircraft could climb even farther. Special high-climbing aircraft are, however, not usually available for meteorological work, and up to the beginning of the last war no useful investigations of the stratosphere had been made by this means. It is now possible to explore up to 13 or 14 km. with a variety of aircraft, into some of which considerable quantities of equipment can be packed. The main advantage of aircraft work is this possibility of carrying a lot of equipment and an observer, and of doing so fairly regularly.

For obvious economic reasons it is normally easier to obtain the use of military rather than civil aircraft for high-altitude research. Unfortunately, however, military aircraft are often very highly powered, resulting in excessive noise and vibration, and there is usually far less space available than in civil aircraft. Both types require a highly developed and efficient servicing organization to keep them in good condition, and one of the important difficulties in this type of work is that of fitting in the maintenance and modification of research equipment under these circumstances.

Technical troubles in aircraft work depend largely on the type of measurement to be undertaken. If the apparatus is robust and insensitive the only real difficulties are those of installation. Although there is spare space in most aircraft, it may be so difficult to reach as to be more or less useless, and it is often quite difficult, for aerodynamic reasons, to place the necessary measuring elements on the outside of the aircraft. With sensitive apparatus in a piston-engined aircraft acute trouble can arise from the intense vibration, which may render the frequency range 30 to 1000 c./sec. unusable if the apparatus can respond to it. Experimental conditions on the ground may also be much worse than laboratory conditions. Aircraft are often left to stand out in the open and are rarely watertight. Simple facilities, such as a soldering iron, may not be available at the apparatus *in situ*.

The actual movement of the aircraft can also create trouble owing to the high forward speeds needed to maintain lift at high altitude. For example, large corrections have to be applied to thermometer readings to allow for the near-adiabatic compression which takes place at the element.†

1.43. *Rockets*

Rocket research has been widely developed in the United States in the last few years, starting from the adaptation of German V 2 rockets and continuing with missiles specially designed for high-altitude research. The outstanding advantage of this costly method of research is the altitude which can be reached—215 km. at present, with no obvious limitations (Greenstein, 1949; Newell, 1950). Nor are weight restrictions severe; the 14 ton, 47 ft. V 2 can carry 2000 lb. in a space of 17 cu.ft. to 160 km. The Aerobee, constructed of non-magnetic materials especially for upper-atmosphere magnetic research, is about 19 ft. long and can carry loads of 150 lb. to 110 km. Little has been published about the U.S. Navy Viking, but it is believed to be a large rocket, like the V 2, although of a more advanced design.

Fig. 4 shows an Aerobee rocket containing a total field strength magnetometer. Since the rocket was especially designed for this kind of work, a reasonable proportion of the space is available for equipment, although this is mainly taken up with the telemetering set (see below) and the power supplies. This rocket has only the arrow stability provided by its large fins and consequently has to be launched up a long vertical ramp with the aid of a booster.

Fig. 5 shows a sectional drawing of a V 2. The particular missile shown was used for a variety of experiments. The spectrograph was for recording solar ultra-violet spectra, the ionosphere trailing antenna for ion-density studies, the cameras were installed in order to record the behaviour of the antenna, the seed-containers were used for a biological mutation experiment, and in the nose of the warhead were a number of Geiger counters for cosmic-ray experiments. In addition, but not shown on the diagram, the rocket contained temperature- and pressure-measuring apparatus.

† For a true adiabatic compression, this correction in °C. is (speed in m.p.h./100)², e.g. 16° C. at 400 m.p.h.

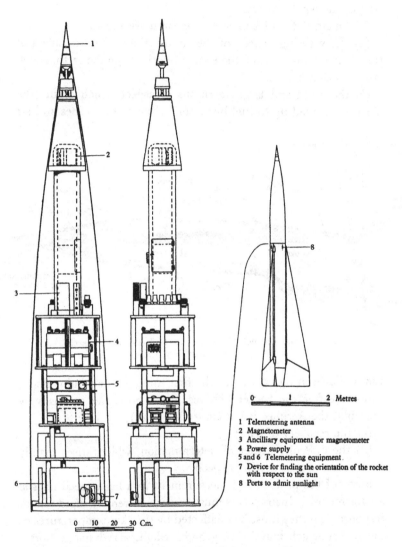

1 Telemetering antenna
2 Magnetometer
3 Ancilliary equipment for magnetometer
4 Power supply
5 and 6 Telemetering equipment.
7 Device for finding the orientation of the rocket
  with respect to the sun
8 Ports to admit sunlight

0   1   2 Metres

0   10   20   30 Cm.

Fig. 4. The Aerobee rocket equipped for magnetic measurements.
(After Maple Bowen and Singer, 1950.)

In comparison with the Aerobee it is noticeable that the space available for instruments (mainly in the warhead) is only a small part of the whole rocket.

The main difficulties in rocket research arise from

(a) the very high speed of the rocket (the maximum speed of the V2 is 1·5 km./sec.; cf. the escape velocity from the earth, which is 10 km./sec.);

(b) the rapid and large variations of aspect which occur (the V2 is controlled up to fuel burn-out by the carbon vanes and air

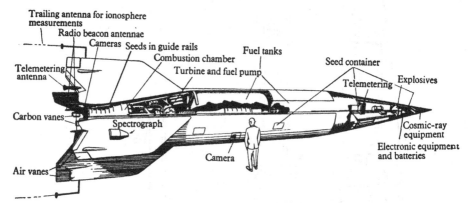

Fig. 5. The V2 rocket equipped for high-altitude research.
(After Newell and Siry, 1946.)

vanes shown in fig. 5, i.e. for the first minute of a 7-min. flight, and thereafter the rocket behaves like a projectile and is very unstable; the Aerobee is not controlled at all);

(c) the recovery of data;

(d) the difficulty of height determination, which arises mainly on account of the very high speed.

Some of the problems of measurement in a high-speed missile are formidable. Instruments must have an extremely rapid rate of response, and they must be unaffected by the skin temperatures of the rocket which may be very high. Slight aerodynamic flutter often leads to violent body vibrations which will severely test any delicate equipment. For some experiments, where the apparatus is simple and sturdy, these factors are not important. For others, however, they place limitations upon the accuracy obtainable, and the most satisfactory solution would be to drop the instruments concerned from a high altitude by parachute. It is known that the Americans have tried to do this and that the Germans intended to do the same, but apparently there are difficulties in execution which have so far prevented useful results from being obtained, and the Americans have concentrated their efforts upon measurements made in the rocket itself.

Large variations in the rocket aspect have limited the accuracy of some measurements from V2 and Aerobee rockets. Attempts to measure the aspect during flight (for example, in order to apply corrections to the static pressure measurements, as described in the next chapter) have so far proved unexpectedly difficult. A satisfactory, but expensive, method of solving this problem is to stabilize the rocket throughout flight. It is believed that the Viking is stabilized in this way, although full details of the rocket construction have not yet been released.

The earliest attempts to recover data were aimed at finding a part of the rocket which survives impact at the earth's surface, and arranging that the data should be stored in that part. This method will presumably continue to be used in the future where the record contains a mass of detail, such as a photograph of a spectrum. In order to increase the probability of finding part of the rocket it is now usual to destroy the streamlining by blowing off with explosives (see fig. 5) some part of the rocket, e.g. the bulkhead, on the descent, in which case aerodynamic forces will tear the rocket to pieces and the parts will strike the earth at about 0·1 km./sec. without completely burying themselves. Despite this precaution it is not always easy to find the shattered rocket after impact. Radar and telescopic tracking of the missile in flight, together with sound ranging of the crash of impact, lead to an estimate of the point of impact of the main part of the rocket. Thereafter follows a detailed search which usually discovers a considerable number of parts.

Owing to these hazards in the recovery of data, it is now usual, where possible, for instrument readings to be transmitted to the ground during flight by radio waves (telemetering). The V2 has been equipped with a 20-channel telemetering set in which the instrument readings are converted into pulses of variable length, data being recorded with an accuracy of 1 %. Such apparatus is fairly heavy, and the more modestly sized Aerobee can only carry a 6-channel frequency-modulated set (i.e. the instrument readings are converted into a variable frequency) with an accuracy of 5 %. In both cases the channels may be subdivided by using a multiple switch.

The problem of height determination will probably be solved as techniques improve, although at the moment this limits the

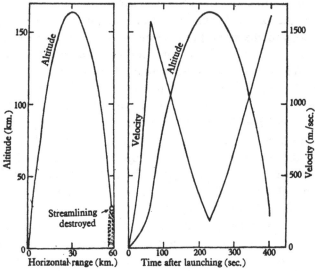

Fig. 6. A V2 rocket trajectory. (After Greenstein, 1949.) The correct ordinate for each curve is marked upon it.

accuracy of many types of measurement. The V2 can carry so much weight that a homing transmitter may be taken up (see the radio beacon antennae in fig. 5), but radar and visual tracking have proved rather inadequate. Most flight paths have been computed from vacuum trajectories with minor corrections, using a few spot readings of position to determine the necessary constants. It is believed that this method can lead to results accurate to 1 km.

The trajectory shown in fig. 6 has been computed by this method. In this case the maximum height reached was 165 km., and the rocket disintegrated on the descent at 30 km. The velocity curve shows how fuel burn-out occurred after 60 sec. at a height of 30 km., and how, thereafter, the missile behaved like a projectile in a vacuum.

CHAPTER II

TEMPERATURE

2.1. Introduction

In recent years the vertical temperature structure of the atmosphere up to 20 km. has been explored regularly by radio sondes from many widely separated points on the earth's surface. Since the last war, soundings up to 30 km. have become more and more frequent.

Above 30 km. the picture is quite different. No longer is it possible or desirable to measure temperature by means of the thermal equilibrium between the atmosphere and a thermometric body, but, instead, a wide variety of phenomena have to be studied, each of which (with the exception of the rocket measurements) leads to information only over a limited height range. Fig. 7 shows the approximate temperature structure of the stratosphere in temperate latitudes, together with an indication of the sources of information which have been used.

None of the measurements made at present above 30 km. is conclusive. They do, however, combine to give a consistent picture below the stratopause, and they are of sufficient intrinsic interest to warrant discussion. But it may be stated at the outset that detailed discussion adds very little to the picture presented in fig. 7.

The importance of an exact knowledge of the thermal structure of the atmosphere is illustrated by the number of physical quantities which may be expressed in terms of it. The relations are particularly simple if the atmosphere can be considered to have the same mean molecular mass at all heights. The next chapter will show that this is a good assumption up to 60 or 70 km., and possibly a little higher.

The velocity of propagation of a sound wave whose pressure fluctuations are small compared with the ambient pressure is given by

$$V = \sqrt{\frac{\gamma k \theta}{m}},$$

where k is Boltzmann's constant, θ is the absolute temperature, m is the mean molecular mass, and γ is the ratio of the specific heats of air. Taking the mean molecular weight for dry air of ground-level composition to be 28·97 (Chapman, 1936), the gas constant per mole to be $8·314 \times 10^7$ ergs deg.$^{-1}$ mole^{-1} (Birge, 1941), and γ to be 1·405 (Jeans, 1921), we obtain

$$V = 2·008 \times 10^3 \theta^{\frac{1}{2}} \text{ cm.sec.}^{-1}.$$

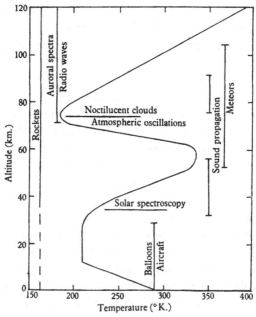

Fig. 7. Methods of measuring atmospheric temperature. (N.B. the temperatures shown are approximate. The results from each different method are given in the text.)

The 'scale height', or the equivalent height of a homogeneous, isothermal atmosphere, frequently occurs in atmospheric calculations and is given by

$$H = \frac{k\theta}{mg} = 2·926 \times 10^3 \theta \text{ cm.},$$

where g is the acceleration due to gravity ($= 980·7$ cm.sec.$^{-2}$; Birge, 1941).†

† Ground-level value. At 100 km. the value should be 3% lower.

Pressure (p) and height (h) are related by the hydrostatic equation

$$\frac{dp}{p} = -\frac{dh}{H}.$$

This may be integrated to give the pressure at all heights in terms of one known pressure,

$$\log_e p(h) = \log_e p(o) - \frac{1}{2 \cdot 926 \times 10^3} \int_0^h \frac{dh}{\theta}.$$

In terms of pressure and temperature, the density is given by the gas laws

$$\rho = p/\theta \times 3 \cdot 485 \times 10^{-7} \, \text{g.cm.}^{-3},$$

where p is in dyne cm.$^{-2}$.

The root mean square and average velocities for an 'average' air molecule may be given in terms of the velocity of sound. These are respectively

$$c = \sqrt{\frac{3k\theta}{m}} = 1 \cdot 46 \, V \, \text{cm.sec.}^{-1}$$

and

$$\bar{v} = \sqrt{\frac{8k\theta}{\pi m}} = 1 \cdot 35 \, V \, \text{cm.sec.}^{-1}.$$

Finally, the mean free paths for air molecules and electrons are given respectively by

$$l_{\text{air}} = \frac{1}{5 \cdot 66 \, \pi r^2 N} \, \text{cm.}$$

and

$$l_e = \frac{1}{\pi r^2 N} \, \text{cm.},$$

where $r = 1 \cdot 87 \times 10^{-8}$ cm. is the average radius of the molecules (Jeans, 1921) and N is the number per cm.^{3} given by

$$N = \rho/m = 2 \cdot 08 \times 10^{22} \times \rho.$$

The corresponding collision frequencies are then given by

$$\nu = \frac{\bar{v}}{l} \, \text{sec.}^{-1}.$$

All the numerical values given above apply to dry air. If air is saturated with water vapour at temperatures near $300°$ K., the effect is to lower the mean molecular weight by a small percentage. In the stratosphere this effect is negligible.

2.2. The lower stratosphere

The discussion in this section will mainly cover the region from the tropopause up to 20 km., which is the only extensively observed region of the stratosphere owing to its accessibility by balloon-borne radio sondes. A vast body of data exists upon this region, but it would not be appropriate here to give it the detailed attention which it merits. Instead, only the salient features will be mentioned, although it should be realized that they are observationally established with a far higher degree of certainty than any other measurements discussed in this monograph.

The position of the tropopause, and therefore the lower boundary temperature of the stratosphere, is sometimes difficult to assess. The British Meteorological Office have been driven to a series of arbitrary definitions of tropopause height (H_t) in terms of changes in the temperature gradient.

'Type I. When the stratosphere commences with an abrupt change in lapse rate to inversion, H_t is the height of the base of the inversion.

'Type II. With an abrupt change to a lapse rate less than 2° K./km., or to isothermal, H_t is the height of the abrupt change.

'Type III. Where there is no abrupt change of lapse rate, H_t is taken to be the height of the point at which lapse rate first becomes less than 2° K./km., provided that it does not exceed this value for any subsequent kilometre.'

These definitions are complex and, to some extent, circular, but nevertheless it is usually possible in practice to assign a tropopause height which seems to represent the position of some kind of thermal transition. Fortunately, the very indefinite type III tropopause is rather rare. On occasion the temperature structure near to the tropopause is such as to defy any precise analysis, and sometimes the definitions allow the existence of more than one tropopause; the term 'multiple tropopause' is then used.

The conditions in the lower stratosphere, below 20 km., vary from place to place and from hour to hour. By taking suitable averages these variations have been related partly to season and partly to latitude. The remaining variations may be termed 'short period' and are usually discussed in relation to weather conditions.

The possibility of referring to a latitude variation of tropopause height is demonstrated by the chart shown in fig. 8, which is one of a series of seasonal charts prepared by Dewar and Sawyer (1947). In the southern hemisphere contour lines are very nearly parallel to lines of latitude, and a similar state is found in the northern hemisphere, apart from distortion over the Asiatic and North American continents. The tropopause height reaches a maximum of 17 km. over the tropics, falling to the north and to the south to an average height of 11 km. in temperate latitudes, and even lower in the less well-observed polar regions.

Fig. 9 shows mean isotherms on a meridional cross-section of the atmosphere. This reveals the interesting fact that not only is the tropopause highest in the tropics, but it is also coldest, so that the tropopause temperature varies in the opposite sense to ground temperature. The figure also shows that the lower stratosphere is only approximately isothermal in temperate latitudes. While the gradient is always numerically smaller than that occurring in the troposphere, it is nevertheless definitely positive in low latitudes and negative in high latitudes. It will be seen that data are missing for the Arctic regions. Here the position is obscure owing to the scarcity of observations and the large seasonal variations which take place. It is possible that during winter there is no obvious tropopause to be observed.

The seasonal change in temperature in the troposphere and in the lower stratosphere up to 20 km. is illustrated in fig. 10. Agra represents a tropical station; the amplitude of the variation of temperature in the lower stratosphere is smaller than the amplitude in the troposphere, and the two variations are out of phase. In fig. 10(b) temperatures for 5–7 km. are compared with those above the tropopause for a number of stations in middle latitudes. The difference from the Agra records is striking; the amplitude of the temperature variation in the lower stratosphere has slightly increased, while the phase has advanced nearly 5 months. The records for Absiko, on the border of the Arctic, show that this trend continues, and that here the stratosphere temperature varies more than the troposphere temperature and lags only one month behind it.

The temperature gradient in the lower stratosphere also shows

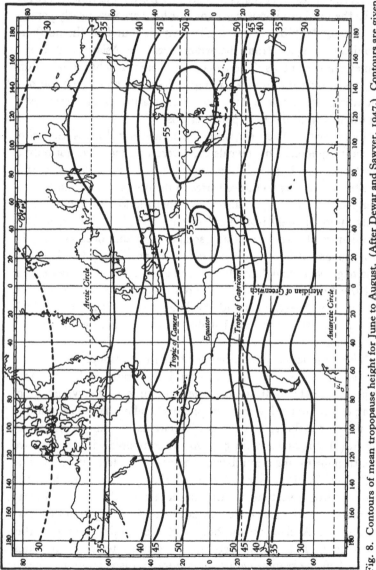

Fig. 8. Contours of mean tropopause height for June to August. (After Dewar and Sawyer, 1947.) Contours are given in thousands of feet. 55 = 16·8 km.; 50 = 15·2 km.; 45 = 13·7 km.; 40 = 12·2 km.; 35 = 10·7 km.; 30 = 9·1 km.

seasonal variations, as can be seen from fig. 9. For example, over England there is a very slight lapse in summer and a slight inversion of temperature in the winter.

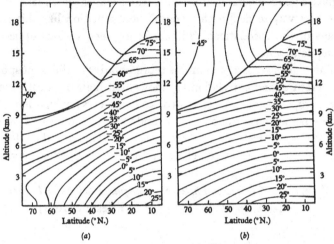

Fig. 9. Mean meridional cross-sections of the atmosphere for the northern hemisphere. (After Wexler, 1950.) The thick line gives the mean tropopause height and the thinner lines are isotherms in degrees C. (a) January, (b) July.

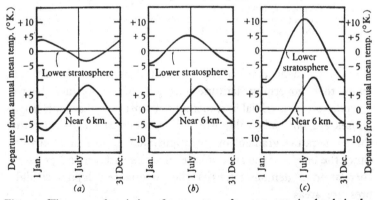

Fig. 10. The seasonal variation of temperature for representative levels in the troposphere and in the lower stratosphere, averaged over several years. (a) Agra, $27°$N.; (b) average over latitudes $43°$N. to $68°$N.; (c) Absiko, $68·4°$N. (After Möller, 1938; and Dobson, Brewer and Cwilong, 1946.)

Superimposed upon the regular latitude and seasonal variations are larger short-period variations which may cause the tropopause height to vary as much as 2 km. in 24 hr. These short-period changes have been interpreted in terms of the horizontal transfer of air masses with all the upper-air properties of some other latitude (Goldie, 1923). Although this explanation is certainly too simple, it gives a good general impression of the changes which take place. The best way to deal with a complicated situation like this is by means of the correlations which exist between various upper-air elements (Table II).

The temperature at 500 mb. is representative of the troposphere temperature, and correlation (b) shows that high tropopauses and warm tropospheres are associated, and vice versa, which is in the same sense as the latitude variation and therefore confirms Goldie's hypothesis. Correlation (d) shows that, by removing the effect of seasonal variations, the correlation (b) is increased, which suggests

TABLE II. *Correlation between tropopause height and other meteorological elements.* (After Priestley, 1944)

Element	Correlation coefficient
(a) Height of the 1000 mb. pressure level	+0·49
(b) Temperature at the 500 mb. pressure level	+0·65
(c) Tropopause temperature	−0·47
(d) Excess temperature at the 500 mb. level over the mean for the time of year	+0·79

that the mechanism governing seasonal changes is quite different from that governing latitude changes of tropopause height. Correlation (c) is trivial and merely demonstrates that high tropopauses are normally cold, which is another way of stating that the temperature gradient in the troposphere is normally negative. Since the height of the 1000 mb. level varies with ground pressure, correlation (a) demonstrates that the tropopause is high over high-pressure areas.

To recapitulate, it may be said that the salient observational facts about the lowest few kilometres of the stratosphere are as follows:

(a) The temperature gradient differs greatly from that observed in the troposphere, being positive in the tropics, almost isothermal in middle latitudes, and probably negative at high latitudes.

(*b*) The tropopause is highest and coldest at low latitudes and lowest and warmest at high latitudes.

(*c*) The seasonal temperature wave changes phase at the tropopause in all latitudes except the polar regions, and the amplitude in the stratosphere varies from smaller to greater than the tropospheric amplitude going from tropics to poles.

The only systematic balloon observations covering the whole of the lower stratosphere have been made by Scrase (1951) from Downham Market, Norfolk, England (52° 36′ N.), and Lerwick,

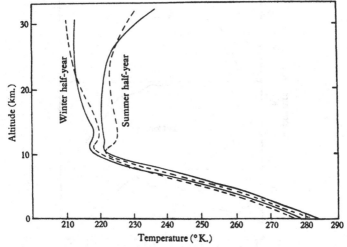

Fig. 11. Temperature over the British Isles, up to 30 km. (After Scrase, 1951.) Averages for summer and winter half-years, 22.00 G.M.T. ——— Downham Market (Norfolk); - - - - Lerwick (Shetland Islands).

Shetland Islands, Scotland (60° 08′ N.), and by Brasefield (1950) from Belmar, New Jersey, U.S.A. (40° 12′ N.). Scrase has obtained sufficient results to break down into summer and winter half-years; the results are shown in fig. 11.

The process of averaging employed in fig. 11 gives a slightly distorted picture, particularly near the tropopause, where the sharpness of the transition is obscured. Of particular interest in fig. 11 is the beginning of a large temperature inversion at 25 km. during the summer half-year. The presence of this inversion has been known for some time from indirect evidence, but the results shown here are the first to demonstrate its existence by direct thermometry.

Inspection of fig. 11 shows that there are very large seasonal changes of temperature near 30 km. This is shown more clearly in fig. 12, which is prepared from the same data. The seasonal range of temperature increases markedly with height through the lower stratosphere, until at 30 km. it is 35° C., with its maximum at the

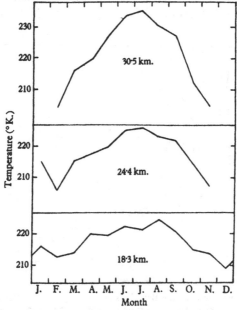

Fig. 12. Seasonal variation of temperature in the lower stratosphere over the British Isles. (After Scrase, 1951.)

summer solstice. This immediately suggests that the direct effect of insolation upon atmospheric temperatures increases rapidly throughout the stratosphere, and it will be seen in Chapter VI that this is so.

The latitude variation of stratosphere temperature at 30 km. also differs very considerably from that near the tropopause. A preliminary estimate is now available from the work of Scrase and three other observers and is shown in Table III.

The results shown in Table III are mainly from balloon measurements and partly from speed-of-sound measurements (see the next section). The temperature maximum for both summer and winter lies about 40° north of the equator instead of near the poles as for the tropopause temperature.

It is perhaps appropriate to add here a warning that it may be some time before regular and direct measurements are made throughout the lower stratosphere. One reason for this is that daylight measurements may be very inaccurate owing to the direct absorption of solar radiation by the thermometer element. Radiation shields are always used on radio sondes, but at high levels there is virtually no ventilation, and thermometer and shield act as a thermal unit. Brasefield (1948) has demonstrated that such shielded thermometers may read 20° C. too high at 30 km. The best procedure appears to be to make the thermometer element

TABLE III. *Temperature at 30·5 km.*

Latitude (° N.)	Winter (° K.)	Summer (° K.)
65	205	221
60	210	229
52	212	232
40	230·5	235
32	224·5	233
9	—	219·5

with the highest possible reflectivity and then to expose it freely to the atmosphere. This also helps to minimize another possible effect, namely, that the thermometer may come into radiative equilibrium rather than conductive equilibrium with the atmosphere. These do not necessarily lead to the same temperature. The measurements of Scrase, which have been quoted, were made at night, thus eliminating the solar radiation error but possibly not the error arising from this second source.

Possible insolation errors during daytime make it very difficult to discuss the diurnal variation of stratosphere temperature. This is a matter of very great importance, since it could give a measure of the degree of direct solar control over atmospheric temperature. Diurnal variations of a few degrees centigrade have been reported, but it is safe to say that no systematic variation has yet been properly established.

2.3. Propagation of sound

Since the velocity of compression waves is proportional only to the square root of temperature, a very attractive method of temperature measurement is suggested which avoids all the difficulties of radiation errors. Equipment has been designed which uses this principle by sending aloft on a balloon a source and receiver of sound together with some means of determining the time taken for the waves to traverse a short path (Barret and Suomi, 1949). Modern electronic techniques could make this fairly easy, were it not for more fundamental difficulties. At low pressures it is found that the efficiency of the source decreases, and at 10 mb. pressure 200 times the energy has to be used as at 1000 mb. pressure. This leads to difficulties with Laplace's theory of propagation, which has to be profoundly modified if the compression is not very much smaller than the ambient pressure. At 50 km. the ambient pressure is about 700 microbars, while at 100 km. it is about 0·5 microbar. The minimum usable source strength has yet to be determined, but it seems likely to be larger than the 100 km. pressure.

It has also been suggested that the frequency of the sound waves might be too high to allow equipartition of energy between the translational, vibrational and rotational modes of the molecules, thus leading to uncertainty in the value of the ratio of the specific heats of air. Below 30 km., however, difficulties of this nature will only arise if the frequency is above 5×10^5 c./sec. (Barret and Suomi, 1949), and far lower frequencies have to be employed for quite different reasons (see below).

For various reasons these direct measurements of sound velocity have never been carried out, and instead experimenters have concentrated upon the very complicated problem of the refraction of waves emitted from ground-level sources, and upon such measurements is based most of our knowledge of the temperature of the atmosphere from 30 to 60 km.

The abnormal propagation of sound waves was first noted in 1901 when the minute guns fired at Queen Victoria's funeral were heard far to the north of London. Interest became focused upon the problem during the 1914–18 war, when gunfire from the western front was frequently heard in southern England.

The term 'abnormal' is used in connexion with this phenomenon for two reasons. First, there is no correlation between the
intensity of the sound received at great distances and the ground-
level meteorological conditions. Secondly, intensity does not fall
off steadily with distance from the source, but instead there may
be an approximately circular zone of audibility out to perhaps
100 km., a zone of silence from perhaps 100–200 km., and thereafter another zone of audibility. This is well illustrated in fig. 13,
which is a map of the positions at which was heard an explosion
at Oppau, Germany, on 21 September 1921.

The existence of zones of silence clearly establishes this as a
refraction phenomenon requiring the velocity of sound to increase
with height somewhere in the upper atmosphere. Three possible
causes exist:

(a) wind velocity changing with height,
(b) atmospheric composition changing with height,
(c) atmospheric temperature changing with height.

The first is a vector condition and could not account for circular
zones of audibility, although it could account for the distortion
of the zones which is frequently observed. Almost from the
beginning (b) has been rejected as an explanation of the circular
zones, although diffusive separation, leading to an increased
proportion of the lighter constituents at great heights, would
produce the observed effect. Recent work, however, which will be
described in the next chapter, shows that the earlier investigators
were correct. The third alternative (c) was first worked out by
F. J. W. Whipple (1923), and this has been the basis of all subsequent work, with due allowance for the effect of winds.

Many investigations have been carried out since Whipple's
original work, and, as an example, we will briefly describe the
apparatus used by an American team at the destruction of the
Heligoland fortifications in April 1947 (Cox, Atansoff, Snavely,
Beecher and Brown, 1949).

The apparatus was designed to record microbarographic waves
in the frequency range 0·025–4 c./sec. Such low frequencies have
to be used because the absorption in the upper atmosphere increases very rapidly with frequency. Fig. 14 shows curves due to
Schrödinger (1917), giving lines of constant absorptivity over a

kilometre path at various heights in an isothermal atmosphere. The figure shows at a glance that, for compression waves to be transmitted to 100 km. and back, the frequency must certainly be less

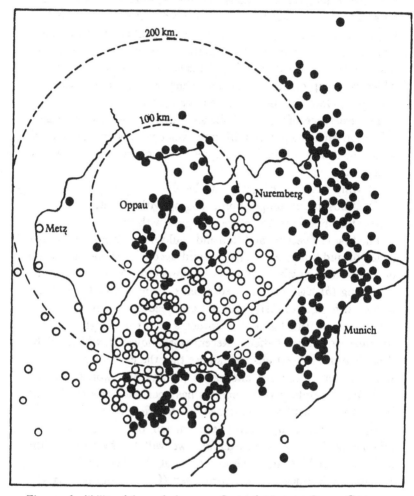

Fig. 13. Audibility of the explosion on 21 September 1921 at Oppau, Germany. The circles represent observers who did not hear the explosion, while the dots represent those who did. (After Mitra, 1948.)

than 10 c./sec. This arises from the rapid dissipation of energy which takes place from compression to rarefaction if the molecular mean free path approaches the length of the wave itself.

The detector was a membrane over a closed cavity coupled mechanically to the movable yoke of an inductance. A second fixed inductance was connected to form a bridge with the first and the circuit was fed from a 10 kc./sec. oscillator. The out-of-balance signal from the bridge was amplified, rectified, and recorded by stylus. The upper frequency limit was imposed by the recorder system, while the lower was imposed by an adjustable leak in the cavity which tended to equalize the pressure on the two sides of the membrane.

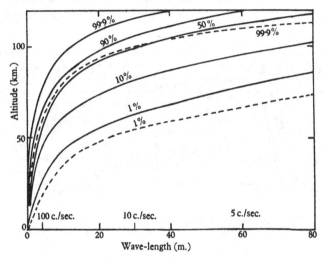

Fig. 14. Percentage energy absorbed per kilometre path length in an isothermal atmosphere at o and −45° C. (After Schrödinger, 1917.) —— 0° C.; ---- −45° C.

Fig. 15 shows a recording made with this apparatus from a relatively small explosion. The abnormal signal, arriving 77 sec. after the normal signal (or ground ray), has an amplitude of 220 microbars and is as intense as the normal signal.

The observations by Cox and others (1949) at the Heligoland explosion are unique as regards the magnitude of the explosion, and they claimed to have recorded reflexions from the inversion above the stratopause as well as from the inversion from 30 to 60 km. Unfortunately, these results are open to question, since it was assumed that the wind speed was unimportant at all levels on the basis of radio-sonde observations up to 20 km., while, as we shall see in Chapter V, it is possible for winds to be strong at 50 km. even though they are weak at 20 km. In fact, it requires a very well-designed experiment to separate the effects of wind and temperature variations.

Crary (1950) has performed experiments on a large scale in Alaska (summer and winter), Bermuda (summer only) and the Panama Canal zone (summer only). He obtained readings of the travel time for explosions from several different points of the compass and for a number of distances between the source and the recorder. By using an array of microphones at the recording point

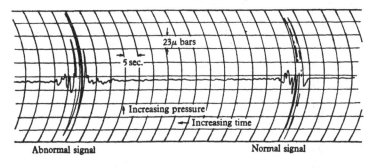

Fig. 15. Two microbarographic records received 182 km. distant from a single explosion of 125 tons of T.N.T. (After Cox and others, 1949.)

he also obtained the horizontal velocity component of the arriving wave and hence the direction of the wave normal by comparing the result with the ground-level velocity obtained from a separate small explosion near the recorder. All these data were obtained in the minimum possible time either by moving both the source and the recorder over land, as in the Alaska experiments, or by dropping rows of 500 lb. bombs from an aircraft over a range between 100 and 500 km. from the fixed recording stations, as in the Bermuda and Canal zone experiments.

To make it possible to separate the wind and temperature effects Crary had to assume that data collected over several days referred to the same wind and temperature structure. The first step was to prepare curves showing the wave velocity as a function of height for each azimuth. For a horizontally stratified medium the wave velocity is proportional to the sine of the angle of incidence, even though it may be caused partly by the velocity with respect to the medium and partly by the motion of the medium itself. The wave velocity and the angle of incidence were measured at ground level, so that the relation between wave velocity and altitude implies a relation between angle of incidence and altitude, leading to the whole ray trajectory. The trajectory could be determined with certainty up to 20 km. from radio-sonde data, and above this level various possible relations were tried until the correct travel times were obtained for all ranges.

At any one altitude the velocity data obtained in this way could be plotted against azimuth, enabling the magnitude and phase of the sinusoidal component caused by steady winds to be picked out. The remaining steady component could then be attributed to the velocity with respect to the medium, from which the temperature can be calculated if the atmosphere is assumed to be of ground-level composition.

The results shown in fig. 16 are subject to fairly large errors, but on all occasions there is a definite increase of temperature with height between 30 and 60 km. with maximum temperatures approximately equal to those at ground level. The temperatures shown are slightly lower than those usually obtained from sound-propagation experiments which do not take proper account of wind velocities. Large seasonal and latitudinal variations of temperature at all heights are demonstrated, but the results are insufficient for any detailed study. Discussion of the wind results will be deferred until Chapter V.

Crary has also studied the possibility of diurnal variations of wind and temperature between 12·5 and 60 km. by means of a series of tests, made in an east-west direction only, at New Mexico. The average velocity of the abnormal waves (i.e. the horizontal distance divided by the travel time for that part of the trajectory above 12·5 km.) and the horizontal component of the velocity were examined for any correlation with the time of day, and a negative result was obtained. Unless there is some amazing correlation between the diurnal variations of wind and temperature, this result indicates that they are both small up to 60 km.

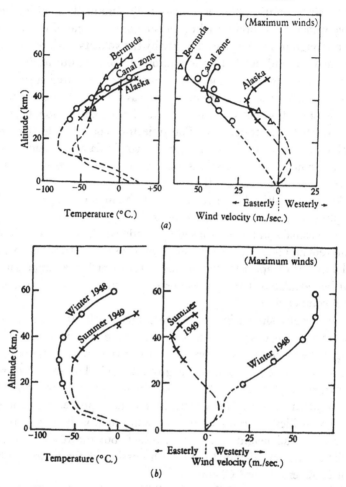

Fig. 16. Winds and temperatures from sound-propagation experiments. (After Crary, 1950.) (a) Comparison of summer conditions in the Canal zone (8° 55' N.), Bermuda (32° 31' N.) and Alaska (64° N.). (b) Comparison of summer and winter conditions in Alaska.

2.4. Meteors

2.41. Introduction

It was a brilliant analysis by Lindemann and Dobson (1923) of the causes of the appearance and disappearance of meteors that first drew attention to the fact that the isothermal structure of the lower stratosphere could not extend much farther than the limits

of balloon observations. These writers also advanced tentative
explanations of their results which form to-day the main basis of
theoretical work on the stratosphere. We know so little about
meteors that air density can be estimated only to orders of magnitude, but Lindemann and Dobson could show that their results
differed from those predicted for an isothermal atmosphere at
220° K. by factors of 10^3 in the region of 100 km.

The success of this early work stimulated efforts to obtain
more accurate meteor observations and a more precise theory.
F. L. Whipple (Whipple, 1943; and Whipple, Jacchia and Kopal,
1949) has been particularly active in this field, and his most recent
results will be described in this section.

Meteor observations do not give very precise values for the
stratosphere temperature. Although they point towards the structure shown in fig. 7, Whipple has nevertheless remarked that
'a constant atmospheric temperature of about 256° K. from 60 to
110 km. is not entirely outside the range of solution'. From the
most recent work it appears that this technique may show itself
to greatest advantage when used to study seasonal variations of
temperature in the upper stratosphere.

A typical meteor, with which we shall be concerned, is visible
to the naked eye. Its apparent brightness may be as great as that
of a first-magnitude star, and it will appear for 1 or 2 sec. somewhere between 120 and 50 km., travelling obliquely with a velocity
of 30 km.sec.$^{-1}$. Its composition will be mainly stone or iron, and
its mass and radius outside the atmosphere will be in the range
10^{-3} to 10^{-2} g. and 10^{-2} to 10^{-1} cm. respectively. It will probably
belong to one of the meteoric showers which form part of the solar
system and which take their names from the constellation away
from which they appear to be directed. It might also be a sporadic
meteor from outside the solar system, but although these are more
numerous than the shower meteors, they are less frequently
observed.

2.42. Technique

Whipple (1943) uses two Schmidt cameras, 37·9 km. apart,
which photograph simultaneously, giving fairly accurate height
determinations by triangulation. A rotating shutter in front of each

camera splits the image twenty times each second, enabling velocities and accelerations to be measured. Luminosity of the trail is measured by comparison with known stars. The data observed are: heights near the beginning and end and at the mid-point of the trail, and also at the point of maximum brightness; luminosity at the same four points, and the integrated luminosity over the whole trail; the velocity and deceleration at the mid-point; the zenith angle of the trail.

The calculation of atmospheric density from these data is performed by means of three fundamental relations, known as the resistance, mass and luminosity equations. These all contain a number of parameters and efficiency factors which can only be evaluated by methods involving considerable speculation.

If the total mass of a meteor is m, then its effective cross-section may be defined as $m^{\frac{2}{3}}A$, where the factor A is taken by Whipple to be unity, after considering the many factors involved. The mass of air encountered in time dt is then

$$dm_a = Am^{\frac{2}{3}}\rho V \, dt,$$

where V is the meteor velocity and ρ is the air density. If the air swept up by the meteor receives a velocity γV, where the dimensionless constant γ is believed to lie between 0·8 and 0·9, then the acceleration of the meteor is given by

$$V' = -\gamma Am^{-\frac{1}{3}}\rho V^2.$$

This is the resistance equation.

The energy lost by the meteor in a given time will be proportional to the product of the air mass encountered and one-half of the square of the velocity. Formally, we may write for the energy taken up as heat by the meteor itself,

$$dW = \tfrac{1}{2}\lambda Am^{\frac{2}{3}}\rho V^3 \, dt,$$

where λ is a coefficient of magnitude about 0·6. If the meteor receives the maximum possible energy, then $\lambda = \gamma$, and therefore the ratio λ/γ is an efficiency factor. Suppose that ζ ergs are required to evaporate or otherwise dissipate 1 g. of the meteor, then the rate of change of mass is given by

$$dm/dt = -\tfrac{1}{2}(\lambda/\zeta)\rho Am^{\frac{2}{3}}V^3,$$

which is the mass equation. The value of ζ depends upon the composition of the meteor and the nature of the process which dissipates its mass. Without discussing the problems in detail we may quote the value used by Whipple, which is $8 \cdot 0 \times 10^{10}$ ergs g.$^{-1}$. To obtain the meteor luminosity, it is assumed that the dissipated matter remains in the meteor coma, and that its energy is partly converted into radiant energy with an efficiency factor τ. The luminosity I is then given by

$$I = -\tfrac{1}{2}\tau(dm/dt)\,V^2,$$

which is the luminosity equation. The parameter τ is particularly difficult to evaluate and the tentative value chosen by Whipple is $10^{-3 \cdot 10}$.

The above three fundamental equations may be combined to give four relations between the measured quantities and the air density. The relations are partly dependent and lead to very similar results; one, however, gives the most consistent results and has been used to obtain those results which are described in the next section. This is obtained by a simple algebraic manipulation of the three fundamental equations,

$$\rho = KV^{-\tfrac{10}{3}}(-V')^{\tfrac{2}{3}}\,I^{\tfrac{1}{3}},$$

where $\qquad K^3 = 4\zeta V/\lambda\gamma^2\tau A^3.$

The important feature of this relation is that I and τ appear raised to the $\tfrac{1}{3}$ power, and although they are both very difficult to estimate, they do not influence the final accuracy very greatly.

2.43. Results

Fig. 17 shows a set of results corrected for seasonal variations (see below); the line shown is not the best fit, but the relation between density and height given by the American Standard N.A.C.A. atmosphere (Warfield, 1947), which corresponds roughly with that shown in fig. 7. The minimum slope near 60 km. is a result of the temperature maximum at this height, and the maximum slope near 85 km. is consistent with a temperature minimum at the stratopause.

Whipple and his collaborators have been able to demonstrate a marked seasonal variation of density in the upper stratosphere by separating their observations into summer, winter and mid-season groups. The density at 78 km. varies through the year by

12 %, a slightly larger variation than occurs at ground level, and the lowest density or maximum temperature occurs in midsummer. If looked upon as a change in level of a constant-pressure surface, this density variation represents a rise of 8·6 km. between summer and winter at high levels. These density changes are well correlated with ground-level temperature and, even more closely, with

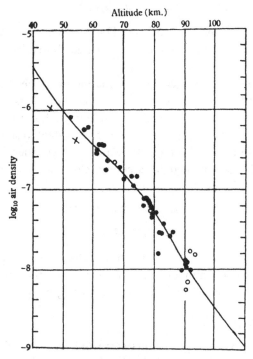

Fig. 17. Density as a function of height from meteor observations. (After Whipple and others 1949.) The crosses are results for which the luminosity varied rapidly, and the circles determinations where the acceleration had a large standard error. The full line corresponds to the relation given by the N.A.C.A. standard atmosphere. All results are corrected for regular seasonal variations.

the normal ground temperature for the time of year (correlation coefficient 0·91), showing that the upper atmosphere does not react to the day-to-day changes at ground level.

Preliminary results from an intensive programme of meteor research in the U.S.A. (Whipple, 1952) show some indication of a latitude variation of temperature in the upper stratosphere. The results shown in fig. 17 were obtained in Massachusetts (43·5° N.).

Observations made near the U.S. rocket-firing range in New Mexico (32° N.) agree more closely with the most recent rocket measurements giving densities and temperatures significantly below the N.A.C.A. standard above 50 km. The best fit to the older measurements has been stated by Whipple to be a flat maximum of 375° K. near 60 km., a rapid drop to 250° K. near 80 km. and a constant or slowly rising temperature from 80 to 110 km. On the other hand, the New Mexico rocket results give temperatures 40–90° K. lower between 50 and 90 km.

2.5. Rockets

2.51. *Methods*

The capture of German V2 rockets in 1945, together with data collected in the hope of performing high-altitude research, enabled the Americans to start what has now become a protracted series of measurements of temperature up to 160 km. (Durand, 1949; Newell, 1950). The methods are always indirect, involving measurements of static air pressure, the speed of sound or the flow round the rocket, usually by means of different kinds of pressure measurement. Such methods are inescapable, since, as has already been mentioned in §2.2, a thermometric body at these heights would probably reach radiative rather than conductive equilibrium with the atmosphere, and even if true temperatures could be reached the compression of air at the element could lead to readings 1000° K. too high.

The three (partly dependent) methods which have been used are:

(*a*) Measurement of static air pressure as a function of height with the subsequent use of the hydrostatic equation in the form

$$\frac{d\log_e p}{dh} = -H = -\frac{mg}{k\theta}.$$

(*b*) Measurement of ratio of the ram pressure at the tip of the rocket to the static pressure, from which, according to supersonic theory, the velocity of sound can be obtained,

$$\frac{\text{ram pressure}}{\text{static pressure}} = 1 \cdot 29 \left(\frac{V_r}{V}\right)^2 = 1 \cdot 29 \frac{mV_r^2}{\gamma k\theta},$$

where V_r is the speed of the rocket and V the speed of sound.

(c) The velocity of sound can also be obtained from an examination of the angle of the shock wave at the projectile's nose, or, as an alternative, by detonating grenades at great heights and timing the travel time of the sound to the ground.

The best results, so far, have been obtained by method (a), and the rest of this section will be concerned with this method alone. The few measurements which have been made by methods (b) and (c) agree well with those made by method (a). When the techniques of the grenade and shock-wave measurements have been further developed they may prove to be more accurate and reliable than method (a).

The instrumental details vary considerably from flight to flight, and fig. 18 is representative of gauge positions only on the earlier

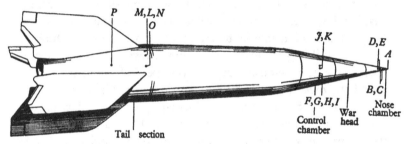

Fig. 18. Pressure and temperature gauge positions on a V2 rocket.
(After Newell and Siry, 1946.)

flights. A is the position of the vent for the ram-pressure gauge, while B, C, D and E are mixed pressure vents and temperature gauges. The temperature measured here is the rocket skin temperature and bears no relation to the air-temperature measurements. F, G, H, I, J and K are more skin-temperature gauges, while L, M, N and O are the static-pressure vents, upon which the most reliable measurements are based. These last vents are placed 15 in. forward of the tail-fin fairing in a position which German workers had found, by wind-tunnel experiments, to give static pressure at all rocket speeds.

The range of pressures to be measured is far too large for any one type of gauge, varying from 665 mm. Hg at the launching site to 10^{-6} mm. Hg at 160 km. Since it is not possible to cover very much more than one decade range accurately with one gauge, this requires a considerable number of gauges. In fact, six ranges have been used, with more than one gauge for each range in order to increase the probability of one working and, if they all operate, to obtain readings from two sides of the rocket to minimize yaw errors. A typical array of gauges is as follows:

Partially evacuated bellows gauges:
760–100 and 100–10 mm.

Pirani hot-wire gauges:
10–1, 1–0·1 and 0·1–0·01 mm.

Phillips ionization gauge:
10^{-3}–10^{-5} mm.

All these gauges are made to produce a potential difference varying with pressure which may be fed into the standard telemetering system. The bellows gauges mechanically operate a very light variable resistor. The Pirani gauges consist of wires heated to 1000° C., the resistances of which vary with the rate of dissipation of heat by means of molecular collisions, and can therefore operate the telemetering system directly. The Phillips gauges utilize a magnetic field to increase the flow of ions set up between two electrodes by a 3000 V. potential difference, and the potential difference caused by the ion current flowing through a resistor is measured.

2.52. Results

In order to check the measurements below 30 km. a pilot balloon carrying a standard radio sonde is normally released just before the rocket flight. The rocket flight is rarely an unqualified success, since a proportion of the gauges always fails to operate and the influence of yaw may be unknown. (Since the ram pressure may be five times the static pressure, and since angles of yaw may be very large, this is a very serious source of error.) Even if the effect of yaw is known, discrepancies sometimes arise which throw doubt upon the results. An encouraging feature of the measurements is the agreement obtained with balloon measurements over the range of altitudes where they coincide and where the rocket is travelling at its maximum speed (see fig. 20). There do not therefore appear to be systematic errors at low altitudes.

Various estimates of error have been given. Newell (1950) states that the pressure measurements are correct to 10% below 70 km. but only within a factor 2 above this level. According to Kellogg and Schilling (1950) the errors in temperature measurement are ± 25° C. from 50 to 60 km., ± 15° C. from 65 to 70 km., ± 20° C. at 72·5 km., and ± 40° C. above 100 km. These are rather subjective estimates of error and do not include possible systematic

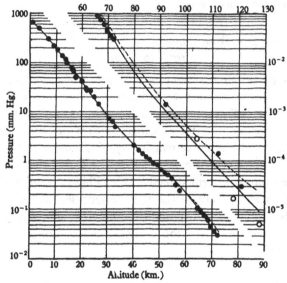

Fig. 19. Air pressure as a function of altitude for two V 2 flights. (After Newell, 1948.) ---- 7 March 1947; —— 22 January 1948; ● Ascent 22 January 1948; ○ Descent 22 January 1948. Experimental points are not given for the flight of 7 March 1947.

errors, but they show that the experiments are accurate enough to delineate the general temperature structure below the stratopause.

Fig. 19 shows the pressure versus height curves obtained on two V 2 flights up to the ionosphere, and the analysis in terms of temperature, together with balloon-sonde data, is shown in fig. 20. The magnitude of the errors involved is shown by the fact that temperature is given by the slope of the curves in fig. 19; a straight line therefore represents an isothermal region. A straight line cannot be fitted to the whole range of results, but fig. 20 could equally well be replaced by two or three isothermal regions.

The American workers have now managed to obtain a large number of results which agree remarkably well (Havens, Koll and LaGow, 1952; Whipple, 1952). They believe that there are real day-to-day changes of temperature which are larger than the errors of measurement. One midnight firing has not indicated any marked difference between day and night temperatures. One firing at the equator gave results similar to those obtained 32° N., and has

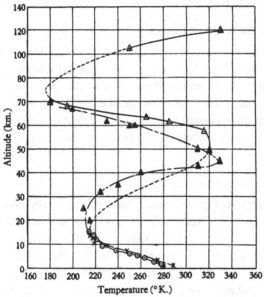

Fig. 20. Air temperature as a function of altitude (from the results shown in fig. 19). V₂ data: △ 7 March 1947; ▲ 22 January 1948. Balloon data: ☉ 7 March 1947; × 22 January 1948.

therefore failed to confirm the latitude variation of upper stratosphere temperature suggested by the meteor results. There is some suggestion of a seasonal increase of pressure from summer to winter between 60 and 70 km., although it is not well established. This result also is at variance with the meteor results already discussed.

The six best firings at White Sands give results which are all compatible with the same temperature distribution, viz. a maximum of 270° K. at 50 km., a minimum of 190° K. at 80 km., followed by an increase to 500° K. at 160 km. This last figure is subject to large random errors together with an uncertainty of about 50° K., depending upon whether oxygen is assumed to be in the molecular or atomic state (the latter assumption leads to the lower temperature and is believed to be correct above 90 km.). It is also possible that the decomposition of oxygen is not the only change in composition which takes place above the stratopause (see §3.41), and this may lead to further errors in the value of the temperature at 160 km.

2.6. Confirmatory evidence

In this section we will discuss briefly various lines of evidence, partly speculative, which add to our general knowledge of the temperature of the stratosphere. These do not carry the same weight as the work described in the previous sections, either because the interpretation is not unique or because the data are sparse. They can therefore, at best, only be considered as confirmatory.

2.61. *Noctilucent clouds*

Humphreys (1933) suggested a theory of the formation of noctilucent clouds (see Chapter III) which leads to a value for the air temperature at 80 km., which is their level of occurrence. Because of their lack of iridescence he assumed that the clouds consist of ice particles, and he also assumed that the concentration of water vapour at 80 km. is one part in 4000 of all the gases present. Whether or not the observed particles are ice is uncertain, but it is a very reasonable speculation. According to measurements made in the lower stratosphere, however, the value of 1 in 4000 for the concentration of water vapour is far too high. There may, moreover, be photochemical destruction of water vapour taking place near the stratopause which might make the concentration at 80 km. still smaller than in the lower stratosphere.

Humphreys's argument states that the ice crystals must be in equilibrium with the water-vapour pressure for condensation to take place, and, for the assumed concentration, this demands a temperature of 160° K. This is a very attractive theory, and, in agreement with other evidence, it demands a low temperature at 80 km. Nevertheless, there are very strong objections to the theory.

Modern evidence suggests a far smaller concentration of water vapour than was assumed by Humphreys, and this requires stratopause temperatures below 160° K. On the other hand, if there is water vapour present at the stratopause, it may be shown, by another theory put forward by Humphreys, that the absorption of radiation will tend to prevent the temperature falling far below 200° K. Thus the theory is not consistent either with the rocket and meteor measurements or with simple considerations about the heat balance of the atmosphere. It is, however, of historical interest, since it was the first to suggest that there might be a minimum of temperature in the upper atmosphere.

2.62. *Solar spectroscopy*

In Chapter IV two methods of measuring the temperature of the ozone layer by spectroscopic methods will be described. The method using the Huggins bands probably refers to a temperature corresponding to the air temperature at 25 km. The more exact method of Adel must refer to a lower level, although the results are scattered around 230° K. It is very important that the theory of these methods should be better understood, since they supply a simple means of measuring a stratosphere temperature from a ground station, and of doing so regularly.

2.63. *Atmospheric oscillations*

Pekeris (1937) first pointed out that the observed free periods of oscillation of the earth's atmosphere can only be adequately explained if the temperature of the stratosphere decreases with height in the region 60–80 km. Wilkes (1949) has reviewed and extended this work in a monograph in this series. Here it is shown how the atmosphere must have two free periods, one to account for the semi-diurnal variations of ground-level pressure, and one to account for the propagation velocity of pulses such as those created by the Krakatoa explosion or the falling of the Great Siberian meteor. The second free period is mainly dependent upon the temperature structure up to 60 km., and computation from the best available data gives good agreement with observation. The first, whose existence can hardly now be doubted, requires a low-temperature region above 60 km. Since the period of this oscillation must be within a few minutes of 12 hr., any proposed temperature structure can be subjected to a rigorous test for consistency by calculating its free periods. A temperature structure found by Wilkes to give good agreement with the observed free periods is similar to fig. 7, except that the stratopause is replaced by an isothermal region from 75 to 110 km.

2.64. *Scale heights and electron-collision frequencies*

In this section we will discuss briefly some of the methods by means of which temperature has been measured above the stratopause. A very large body of data has been collected over the past 15 years by many different workers (see Gerson, 1951, for a general review) using differing techniques, and they all indicate a positive gradient of temperature of the order of 10° K./km. from the stratopause up to 400 km. It will be seen in the next section that this result is not confirmed by spectroscopic measurements.

The two methods which yield reasonably precise results are the determination of scale heights from the vertical distribution of light intensity in the aurorae or from the electron concentration as a function of height, and measurements of electron-collision frequency.

The auroral measurements are not of great accuracy owing to the difficulties of accurate photometry when the source is unsteady, as is generally so with aurorae. Furthermore, a theory of the formation of aurorae is required before the results can be interpreted in terms of scale height, and the theory used is speculative, and may perhaps be in error. Similar remarks apply to the electron-concentration measurements which are based upon the determination of the height of reflexion from the ionized layers as a function of the frequency of the incident radio waves.

The collision frequency can be found from measurements made upon the cross-modulation of two radio channels or from measurements of the reflexion coefficient as a function of the frequency of the incident waves. In order to interpret the results it is necessary to know the collision cross-section between electrons and the various constituents of the upper atmosphere and also the relative concentrations of these constituents.

Thus, apart from the theoretical difficulties with the scale-height measurements, both types of measurement depend upon a knowledge of the composition of the ionosphere. It is too far outside the scope of this monograph to discuss this extremely difficult subject, except to say that our knowledge is most uncertain (see, for, example, Spitzer, 1949) and has been rendered even less certain by the recent measurements of rare-gas concentrations near 70 km. which are described in §3.41. The results displayed in Table IV are therefore not conclusive and may possibly be greatly revised in the next few years.

Radio scale-height measurements have yielded one result in a region of the atmosphere where the composition is probably known. Budden, Radcliffe and Wilkes (1939), making measurements on the lowest sections of the E layer, found a value for the

TABLE IV. *Ionospheric temperatures*

Altitude (km.)	From electron-collision frequency (° K.)	From auroral scale heights (° K.)
100	300	219
150	825	531
200	1350	1580
250	2175	2073
300	2400	2455
350	3285	2704
400	3450	—

scale height of $6 \cdot 0 \pm 0 \cdot 5$ km. at a height of 70 km. Assuming ground-level composition for the atmosphere at this height, this indicates a temperature near the stratopause of $177 \pm 15°$ K.

Gerson prefers the electron-collision frequency and the auroral scale-height data for finding the trend of temperature through the ionosphere. His results for the annual mean temperature in temperate latitudes are shown in Table IV.

2.65. *Auroral and nightglow spectra*

Temperature determinations from auroral and nightglow spectra can be based upon three principles (Swings, 1949; Harang, 1951; Meinel, 1951):

(*a*) measurements of Doppler line widths,

(b) measurements of the distribution of intensity in the rotation lines of a molecular band,

(c) comparison of the intensities of bands with differing vibrational quantum numbers.

Method (a) is the most interesting, since it leads without doubt to the molecular kinetic temperature. The Doppler width of a line is given by

$$\frac{\Delta\lambda}{\lambda} = 7\cdot2 \times 10^{-7}\sqrt{\frac{\theta}{m}},$$

where m is the mass of the emitting molecule, which, unlike the mean atmospheric molecular mass, is known provided that the emission is identified. Unfortunately, a resolving power of about 400,000 is required for effective measurements, and while such resolving powers are possible they have not yet been applied to this work. It is also necessary to observe a line with a very small natural line width, and this eliminates the possibility of working upon lines arising from permitted transitions; fortunately, a number of relatively intense lines in the auroral and nightglow spectra arise from forbidden transitions. Babcock (1923) has made interferometric measurements on the 5577Å. O 1 line with order of interference 85,000. He was only able to establish that the line width was less than 0·035 Å., corresponding to a temperature less than 1200° K. Vegard (1937) has also made interferometric studies upon this line in the auroral spectrum. Since his interference patterns were equally sharp whether they came from parts of the aurora at 100 or at 200 km., he concluded that the temperature difference between these levels is less than 250° K.

Molecular rotational temperatures are obtainable from a study of the shape of molecular band contours or, if the resolving power, of the spectroscope is high enough, from the relative intensities of individual rotation lines. The intensity k_J of a line corresponding to the rotational quantum number J is given by

$$\log_e \frac{k_J}{J} \simeq \frac{h^2}{8\pi^2 k\theta I}J(J+1)+\text{const.},$$

where h is Planck's constant and I the moment of inertia of the molecule (see §6.21). Since the rotational quantum number of a line is known from its position in the spectrum, the slope of the

line obtained by plotting $\log_e(k_J/J)$ against $J(J+1)$ can be used
to determine the rotational temperature.

Fig. 21 shows one of the nitrogen bands in an auroral spectrum,
the R branch of which has been extensively used by Vegard for
temperature determinations. The majority of such measurements
apply to the 100 km. level, and very consistent results near 228° K.
are obtained. There does not appear to be any increase of temperature when the higher levels up to 800 km. are investigated,
nor is there a difference between sunlit aurorae and those in the
earth's shadow.

Measurements on the nightglow spectrum yield similar results.
The Vegard-Kaplan nitrogen bands give temperatures of 230° K.,

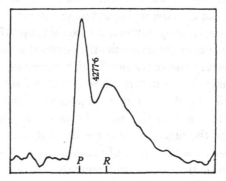

Fig. 21. The contour of the 4278Å. nitrogen band in the
auroral spectrum. (After Harang, 1951.)

while the Herzberg oxygen bands give temperatures of 200° K.
The Kaplan-Meinel oxygen band has given a more precise value
of $150 \pm 10°$ K. and has shown real variations over the range 132–
200° K. from night to night. The level of emission of the Meinel
OH bands can be placed fairly accurately at 70 km., and the
temperature obtained from them is $260 \pm 5°$ K.

When an emission of frequency ν takes place with a change of
vibrational quantum number from v' to v'', the initial population
of the v' level can be found by forming a sum involving the
intensities of all bands with v' in common (Herzberg, 1950, p. 204),

$$N_{v'} = \text{const.} \sum_{v''} \frac{k_{v'v''}}{\nu^4}.$$

If the upper vibrational levels are in thermal equilibrium, then by Boltzmann's law

$$N_{v'} = \text{const. } \exp \left[-\frac{E(v')}{k\theta} \right],$$

where $E(v')$ is the energy of the v' level. From two or more values of $N_{v'}$ therefore a value for the vibrational temperature can be obtained. In this way Rosseland and Steensholt (1932) obtained very approximate values of the temperature in aurorae between 2000° and 7700° K.

It is possible that accurate measurements of the heights of nightglow emissions may show that they all originate near the stratopause, in which case the nightglow rotational temperatures will be consistent with the rocket and meteor results. On the other hand, the rotational temperature measurements up to 800 km. from auroral spectra definitely contradict the results shown in Table IV. There are several possible reasons for this discrepancy. Rotational and vibrational temperatures measured from emission spectra will only necessarily equal the kinetic temperature if the excitation to the upper state is strictly thermal. The excitation mechanisms in the aurora and the nightglow are not thermal, but depend upon inelastic collisions with quanta of radiation or with particles, and it is well known that vibrational temperatures measured in the laboratory depend greatly upon the excitation mechanism (Herman, 1945). While vibrational temperatures are therefore not to be trusted, it is possible that during an inelastic collision the rotational levels maintain the population of the ground state, which should correspond to the kinetic temperature. Nevertheless, the rotational temperatures of flame spectra measured in the laboratory sometimes differ from the kinetic temperature, although when they do so they are usually too high.

In the case of the scale-height and collision-frequency measurements it has already been pointed out that there are theoretical difficulties with the interpretation in terms of temperature, and that the lack of knowledge of ionospheric composition also makes the results uncertain.

At present therefore it is not possible to decide between the results shown in Table IV and the rotational temperatures obtained from auroral spectra. Many writers prefer the former because of other indications of high temperatures in the ionosphere (Spitzer, 1949), but a final conclusion is only likely to be reached when the atmospheric composition can be measured up to and beyond 200 km. and Doppler widths of the forbidden O1 lines in the auroral spectrum can be resolved.

2.66. *Scattering of light*

By observing the scattered light from the zenith sky information can be obtained about the vertical distribution of scattering centres. In the lowest few kilometres of the atmosphere scattering is mainly caused by dust particles and water droplets, which obscure the effect of molecular scattering. When the sun is below the horizon, however, the lower atmosphere is not illuminated and the molecular scattering from the higher layers may be observed. When the sun is 16° below the horizon the layer at 250 km. is just illuminated and the intensity of scattered light is the same as that from the general background of nightglow emissions, zodiacal light and starlight. Levels below 250 km. can therefore be observed by these means.

Molecular scattering follows the Rayleigh law, and therefore the particle density can be calculated if the incident and scattered intensities are known. Since the sunlight may traverse 2000–3000 km. of atmosphere, and since atmospheric attenuation coefficients are not well known, the incident intensity cannot be accurately calculated. Moreover, the reflectivity of clouds and of the earth's surface also enters into calculations of the incident intensity, and these too must be known over distances of 2000–3000 km.

Knowledge of the incident intensity is not so important if measurements are restricted to density gradients. This is the method which has been followed by Link (1934) and by Gauzit and Grandmontagne (1942). The former has been able to fit smooth curves to his derived results of the density gradient as a function of height; indicating that there was no evidence for maxima or minima of temperature. The latter authors have, however, detected discontinuities in the density gradient at 57 ± 7 km., near 95 km. and near 160 km. The first corresponds to a change in temperature gradient of 16·5° K./km., which can be seen to be in reasonable agreement with fig. 7.

These results demonstrate the potential value of this kind of work, although van de Hulst (1949) believes that the results so far obtained are unreliable. Nevertheless, he also believes that by means of a large-scale, well-organized experiment valuable results are possible.

There have been many attempts to measure the scattering from upward-pointing searchlight beams, but only recently has it been possible to obtain searchlights of sufficient intensity or detectors of sufficient sensitivity to give results from levels higher than those normally reached by radio sondes. Elterman (1951) has obtained signals twice as great as the detector noise for light scattered from 62 km., using a photomultiplier with a 60 in. parabolic mirror to view a searchlight 20·5 km. distant. By scanning along the stationary searchlight beam, relative intensities were determined from 15 to 62 km., which were combined with radio-sonde measurements to give densities over most of the stratosphere.

The results obtained by Elterman do little more than confirm in a general way the atmospheric densities computed from the observations which have already been described in this chapter. Nevertheless, this is potentially a most powerful method of measuring stratospheric densities and temperatures. Measurements can be made on any clear night and only relatively simple ground equipment is required. Moreover, the interpretation of the results appears to contain no particular difficulties, in marked contrast to the other methods of temperature measurement which have been described. It may be hoped that a great deal more effort will in future be directed along this line of research.

CHAPTER III

COMPOSITION

3.1. General survey

The chemical composition of the stratosphere differs only slightly from that of the troposphere; nevertheless, the minor differences are of very great importance for two reasons. First, changes in the relative concentrations of the chemically inert constituents are probably indicative of diffusive separation, whereby the proportion of the lighter gases will increase as the height increases. Secondly, the thermal state of the stratosphere is mainly dependent upon the concentrations of three minor constituents, namely, water vapour, ozone and carbon dioxide. Of these three, ozone is unique in that it is the product of a photochemical reaction in the atmosphere itself, and as a result its properties have been studied to an extent which requires a separate chapter to describe.

It must not, however, be assumed that the ionosphere resembles the lower layers in chemical composition, and although this subject does not fall within the scope of this monograph a few words must be said about it. Photochemical theory, which gives a good account of the formation of the ozone layer, indicates that above 90 km. nearly all the oxygen must be in atomic form. Tentative considerations of a similar nature show that water vapour may decompose at levels above 70–80 km., and carbon dioxide above 100 km. There is, moreover, reason to believe that diffusive agencies are dominant somewhere above 70–100 km. (see §3·42), and that the concentrations of lighter components will increase rapidly with height. Finally, auroral and nightglow spectra indicate the presence above the stratopause of oxygen, nitrogen, hydrogen and sodium atoms together with ions and free radicals such as the hydroxyl radical, although in no case need the concentration be large.

As a basis for discussing the composition of the stratosphere we may start with the best available data for the troposphere, which are displayed in Table V.

All the gases listed in Table V, except hydrogen, xenon, radon and carbon monoxide, will be mentioned in this and the next chapter. The vertical distributions of these four gases are unknown, and in the case of hydrogen it is even doubtful whether it exists in the atmosphere, except perhaps as a product of photochemical decomposition of methane and water vapour. Carbon monoxide has only recently been proved to be an atmospheric constituent (see, for example, Bates and Witherspoon, 1952); it should be possible to find its vertical distribution by means of spectroscopic observations similar to those described in §3.52.

TABLE V. *The composition of the troposphere*

Component	Percentage by volume	Density relative to air	Comments
N_2	78·09	0·967	⎫
O_2	20·95	1·105	⎬ Total 99·97% by volume
A	0·93	1·379	⎭
H_2O	$1-10^{-3}$	0·621	Highly variable
CO_2	$2\cdot6 \times 10^{-2}$	1·529	Slightly variable at surface
Ne	$1\cdot8 \times 10^{-3}$	0·695	—
He	$5\cdot24 \times 10^{-4}$	0·138	—
CH_4	$1\cdot6 \times 10^{-4}$	0·558	—
Kr	$\sim 10^{-4}$	2·868	—
H_2	(5×10^{-5})	0·070	Existence doubtful
N_2O	$3\cdot5 \times 10^{-5}$	1·529	Variable at the surface
CO	2×10^{-5}	0·967	Variable
Xe	8×10^{-6}	4·524	—
O_3	$\sim 10^{-6}$	1·624	Highly variable
Rn	6×10^{-18}	7·68	Radio-active, half-life 4 days

In addition to the components listed in Table V, the atmosphere contains solid matter in the form of dust, particles of sea-salt, etc. Few reliable measurements have been made of these, and in any case concentrations are very variable; nevertheless, the general result that the quantity of solid matter falls off very rapidly with height has been well established from solar scattering experiments and direct sampling. Since such particles can be looked upon as a gas of enormous molecular weight, their concentration could be of great interest as an indicator of the intensity of turbulent transfer.

For simple considerations, both troposphere and stratosphere

may be regarded as a mixture of oxygen and nitrogen only, and
the very approximate figures in Table VI give the important data
for this average atmospheric 'gas'.

Of the two main gases nitrogen is the more abundant and the
less reactive constituent, for which reasons it is often used as the
datum against which to measure other gaseous concentrations. It
is, however, of interest to know whether its isotopic ratio varies
with height in view of its observed constancy at ground level.
McQueen (1950) has used a mass spectroscope to measure the
ratio $N_{14}N_{14}$ to $N_{14}N_{15}$ in six samples collected from 40 to 60 km.
during rocket flights. The results were somewhat erratic, although
in five cases it was possible to say that the lighter molecule had

TABLE VI. *The main oxygen-nitrogen mixture up to 100 km.*

Altitude (km.)	Temp. (° K.)	Pressure (mb.)	Molecules per cm.^{3}	Mean free path (cm.)	Collision frequency (sec.$^{-1}$)
0	288	$1 \cdot 013 \times 10^3$	$2 \cdot 5 \times 10^{19}$	$6 \cdot 3 \times 10^{-6}$	$7 \cdot 3 \times 10^9$
11†	218	$2 \cdot 3 \times 10^2$	$7 \cdot 8 \times 10^{18}$	$2 \cdot 1 \times 10^{-5}$	$1 \cdot 9 \times 10^9$
32	218	$8 \cdot 6$	$2 \cdot 9 \times 10^{17}$	$5 \cdot 6 \times 10^{-4}$	$7 \cdot 1 \times 10^7$
62	330	$2 \cdot 0 \times 10^{-1}$	$4 \cdot 5 \times 10^{15}$	$3 \cdot 6 \times 10^{-2}$	$1 \cdot 4 \times 10^6$
84‡	200	$1 \cdot 2 \times 10^{-2}$	$4 \cdot 4 \times 10^{14}$	$3 \cdot 7 \times 10^{-1}$	$1 \cdot 0 \times 10^5$
100	300	$1 \cdot 5 \times 10^{-3}$	$3 \cdot 6 \times 10^{13}$	$4 \cdot 5$	$1 \cdot 0 \times 10^4$

† Tropopause.   ‡ Stratopause.

increased its concentration, the maximum increase being 3·9%.
The validity of these results has been questioned (Hagelbarger,
Loh, Weill, Nichols and Wentzel, 1951), because the helium concentration in the same samples did not show any increase over its
ground-level value.

Oxygen concentrations have been measured up to 30 km. by
means of sampling flasks carried on balloons, and up to 70 km.
using rockets. Since oxygen is a heavy component of the atmosphere its relative concentration might be expected to decrease
with height, and this is, in fact, nearly always observed. It is,
however, almost certain that the main part of the observed oxygen
deficits is caused by chemical reaction with the sampling flask or
with tap grease rather than by any tendency towards diffusive

separation. For what they are worth the results of balloon experiments by four different investigators are given in Table VII.

The rocket measurements of oxygen concentration leave almost no doubt that the oxygen has reacted chemically with the sampling flask. Measurements on samples from 50 to 70 km. show oxygen deficits from 40 to 100 %, and, in the light of results on the rare gases, this cannot be interpreted as the result of diffusive separation.

TABLE VII. *Measured oxygen deficits in the lower stratosphere.* (After Paneth, 1939)

Altitude (km.)	0	9–17	14·5	18·5	19·0	21·5	24	28–29
Percentage deficit	0	0	0·14	0·38	0·24	0·24	1·7	2·5

3.2. Water vapour

3.21. *Measurements before 1940*

The standard humidity-measuring instruments used below 5 km. are the wet-and-dry bulb thermometer and the gold-beater's skin hygrometer, the latter being used on balloon-borne radio sondes. These radio-sonde hygrometers have to absorb a definite quantity of water before they reach equilibrium. For example, Glückauf (1947) states that for a gold-beater's skin to register a change of 10% relative humidity requires the addition or removal of 5×10^{-5} g. of water per cm.^{2} of surface. Since stratospheric air may contain only 3×10^{-10} g./cm.^{3} this demands that a 100 m. column of air must be brought into intimate contact with the film before the change of humidity is registered. Consequently, such hygrometers are extremely slow in action in the stratosphere, and since the balloon is continually rising, very large corrections have to be applied to the readings. Glückauf (1945a) has attempted to apply corrections to observed readings and managed to show that, on occasion, the relative humidity with respect to ice† might be as low as 2–5% in the stratosphere, while the minimum value below the tropopause, which he measured, was 40%. This was the first, although not entirely conclusive, evidence that the stratosphere is comparatively dry.

† Since water and ice at the same temperature have different vapour pressures, except at the triple point, and since in the clean air of the stratosphere either phase may occur, it is necessary to specify whether relative humidity is with respect to ice or water.

3.22. *The frost-point hygrometer*

The study of stratospheric humidity has been revolutionized by the development of the frost-point hygrometer by Dobson and

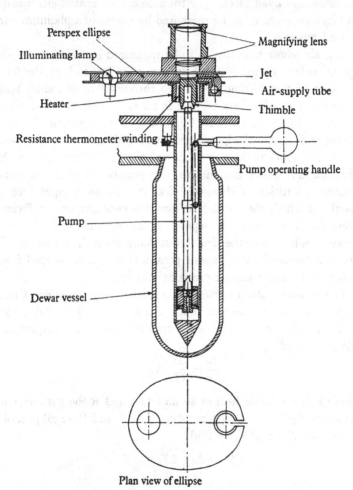

Fig. 22. Manually operated frost-point hygrometer for use on aircraft.
(After Dobson and others, 1946.)

others (1946), the instrument being simply a more sophisticated version of the well-known dew-point hygrometer. Fig. 22 shows the manually operated instrument developed during the last war for use on aircraft.

The pump, shown in fig. 22, is designed to project a jet of liquid air, or petrol cooled by solid carbon dioxide, on to the underside of an aluminium thimble, the upper surface of which is polished, anodized and dyed black. The thimble can be heated electrically, and its temperature can be measured by means of a platinum wire wound round it.

The air whose humidity is to be measured is drawn across the upper surface of the thimble in a narrow jet, and, as the temperature of the thimble is lowered through the saturation temperature, a streak of ice crystals is formed. This streak is given intense oblique illumination by means of an ingenious Perspex ellipse with its outer edge silvered, and with a lamp at one focus and the thimble at the other. Viewed through a lens, against the black background of the thimble, it is possible to detect the most minute quantities of deposited ice. A thimble temperature is sought at which the ice deposit remains constant after sufficient over-cooling to ensure a deposit in the first place. This temperature is known as the 'frost-point temperature', and its significance in terms of water-vapour pressure may be discovered from tables of the water-vapour pressure over ice.

The errors involved in this method are of some interest. Using the Clausius-Clapeyron equation it can be shown that small errors in the vapour pressure (e) and errors in frost-point temperature (T_f) are related by

$$\frac{\Delta e}{e} = \frac{L}{RT_f^2} \Delta T_f,$$

where L is the latent heat of sublimation and R the gas constant for water vapour. Taking $L \simeq 10^4$ cal./g.mol. and $R = 2$ cal./g.mol./ degree, for $T_f = 200°$ K. we find

$$\frac{\Delta e}{e} = \frac{\Delta T_f}{8}.$$

Errors in T_f do not vary very much with the humidity to be measured, and, for the manually operated instrument, $\pm 2°$ K. has been claimed in practice. Thus a more or less constant error in vapour pressure of $\pm 25 \%$ is involved. This is a very small error in comparison with the range of water-vapour pressures which have to be measured, which may be 10^4 to 1, and it is probably

COMPOSITION   61

quite adequate for calculations of the radiative heat balance of the
stratosphere.

The only real difficulty with the frost-point hygrometer is that
below 180° K. ice is deposited on the thimble in the form of an
invisible vitreous layer instead of discrete crystals. Fortunately,
frost-point temperatures as low as this do not seem to occur in the
lower stratosphere.

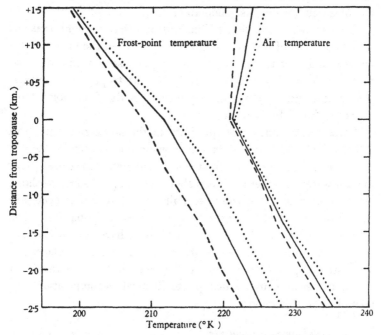

Fig. 23. Average frost-point and air temperatures over southern England.
(After Shellard, 1949.) —— year; ····· summer half-year; ---- winter half-year.

Shellard (1949) has summarized the results of over seventy
high-level ascents in a Mosquito aircraft up to 14 km., where
humidities were measured with a frost-point hygrometer.

Fig. 23 shows Shellard's results for air and frost-point temperatures relative to those at the tropopause level. This method of
plotting is a fairly common device used to allow for variations of
the position and temperature of the tropopause itself. Through
the troposphere, once away from the surface layers, the mean air-

temperature and mean frost-point-temperature curves run nearly parallel. At the tropopause, however, the air temperature, in these cases, begins to increase slowly with height while the frost-point temperature falls off even more rapidly than before. The result is an extremely rapid drop in relative humidity from a mean value of 42·5 % at the tropopause to a mean value of 2·2 % 1·5 km. higher. Sometimes the frost-point temperature falls so rapidly at the tropopause that there appears to be a discontinuity, although the mean curves disguise such an effect.

Further points made by Shellard are that a slight seasonal variation exists of the frost-point temperature in the lower stratosphere, and that the height of the tropopause is well correlated with the frost-point temperature below the tropopause (correlation coefficient, 0·7) but not so well with the frost-point temperature above the tropopause.

These measurements of humidity in the stratosphere confirm Glückauf's far less precise data in showing that the stratosphere is remarkably dry. The lowest frost-point temperature measured so far in the lower stratosphere has been about 190° K. (Dobson and others, 1946), and since this is the lowest air temperature that occurs in the earth's atmosphere (near to the tropical tropopause) it has been suggested that the air concerned must have passed through this region some time before the measurement took place. An alternative explanation might be that the air has recently been above the stratopause, since photochemical decomposition can take place in this region.

3.23. Balloon measurements

The earliest attempts to measure water-vapour concentrations from a balloon, apart from radio-sonde measurements with a gold-beater's skin, were made by Regener (1939–40), using a direct-sampling method. Stratospheric air was drawn through a tube filled with phosphorus pentoxide, and this tube could be weighed before and after the balloon ascent. The change in weight with dry stratospheric air is, however, on the very limits of measurement, for which reason this work does little except confirm in a general way that the stratosphere is comparatively dry.

Recently the apparatus described in the last section has been developed for use in routine balloon ascents (Barret, Herndon and Howard, 1950).

The instrument in its final form is rather complicated and weighs 46 lb. In order to carry such a load to 30 km. a plastic balloon 70 ft. in diameter was required. The equipment consists of (a) an automatic frost-point hygrometer, (b) a thermo-junction for air-temperature measurement, (c) a hypsometer to measure the air pressure, (d) an ice bath as a reference temperature, (e) a very stable d.c. amplifier to raise the levels of the signals from (a), (b) and (c), (f) a standard cell for internal standardization, and (g) a radio telemetering set with a switch to connect the various signals in turn. The telemetering system converts the various thermocouple voltages into pulses modulating a radio-frequency carrier wave in such a way that the pulse frequency varies with the applied voltage.

The hygrometer itself is of particular interest. A lamp illuminates two photoelectric cells, one directly and one by way of a reflexion from the polished surface of a thin sheet of electroplated copper which serves as a thimble. These photoelectric cells are opposed so as to give no signal when a slight deposit of ice is on the thimble. For other conditions the out-of-balance signal is amplified and made to vary the output of a high-frequency oscillator which induces eddy currents in an iron base underneath the thimble surface. At the same time the thimble is cooled by means of a liquefied 'Freon', which is also employed as the hypsometric fluid for pressure measurement. This forms a closed-loop servo-system which will cause the thimble temperature to set itself to the temperature required to hold a small ice deposit. The radio-frequency heating was adopted because the heat is absorbed in the surface layer of the thimble, leading to a high speed of operation. Hunting is prevented and stability improved by means of resistance-capacity couplings in the amplifier, leading to a 'rate of deposit' feed-back, and hence to a damping term.

Fig. 24 shows the results of the three ascents made with this apparatus. In the troposphere the results are much the same as those obtained from aircraft in England. In the stratosphere, however, there are differences, although, since the overlap in height is so small and the geographical positions so different, it is hardly

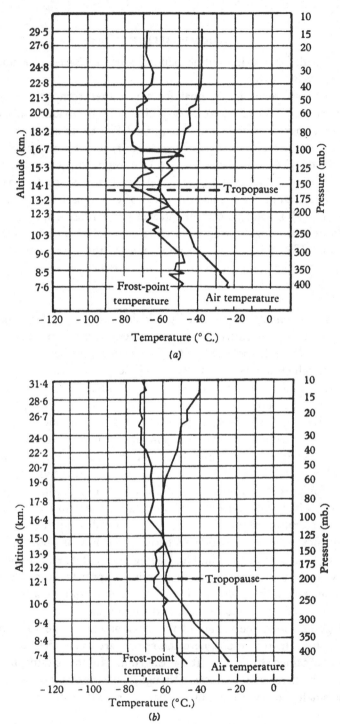

Fig. 24. For description see p. 65.

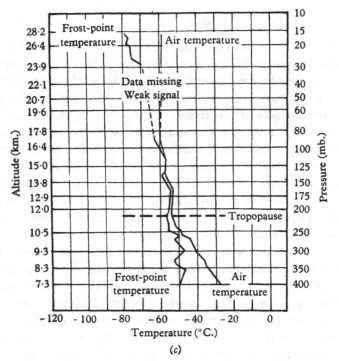

Fig. 24. Measurements of water vapour up to 30 km. (After Barret and others, 1950.) (a) 1 July 1949, Camp Ridley, Minnesota; (b) 26 August 1949, Camp Ridley, Minnesota; (c) 7 January 1950, St Louis, Missouri.

possible to draw conclusions. In general the frost-point temperatures measured in the stratosphere are closer to the air temperatures than the aircraft measurements would suggest. Moreover, the frost-point temperature is very irregular, and it will be seen that on 1 July 1949 there was an apparently supersaturated layer at 16 km.

The ratio of the density of water vapour to that of air does not vary greatly above 14 km. on these three ascents, and, as a rough average, is about 4×10^{-5}. This corresponds to a volume proportion of 6×10^{-5} (cf. Humphreys's assumed value of $2 \cdot 5 \times 10^{-4}$ at the stratopause, as mentioned in §2·61).

3.24. Spectroscopic methods

Water vapour has a number of intense absorption bands in the infra-red spectrum, and it is their existence which accounts for the importance of obtaining an exact knowledge of the concentration of water vapour in the stratosphere, since they affect profoundly the heat balance of this region. The bands may also be used as a means of measuring the amount of water vapour above any level in the atmosphere by taking solar spectra at the level concerned. In fact, the absorption measured in this way may be a more important quantity than the water-vapour concentration, which is often only required to compute it. Owing to the weight and complication of infra-red spectroscopic equipment such experiments have only been made from aircraft.

Of the major water-vapour absorption bands, one is centred at $6 \cdot 3\mu$, and another, the rotation band, stretches from 20μ to the micro-wave region of the spectrum. Difficulties of technique increase with wave-length, and the $6 \cdot 3\mu$ band is therefore the most convenient for identification purposes. Even at these wavelengths it is necessary to use the least efficient of all radiation detectors (thermocouples or bolometers) and to work with amplifications which show the Johnson 'noise' from the detector. It is accordingly necessary to work with very robust electronic equipment and, in order to obtain the maximum radiation intensity, to focus an image of the sun on the slits. This second requirement demands some kind of fast heliostat capable of counteracting the large and fast oscillations of the aircraft.

Solar spectra have been obtained up to 11 km. in the United States from a B 29 bomber adapted for the purpose (Strong, 1949), but the results obtained were not quantitatively interpreted. In this country spectra have been obtained up to 9 km. from a Mosquito aircraft (Goody, 1950a; Yarnell and Goody, 1952) and some attempt has been made at interpretation.

Fig. 25 well illustrates the dryness of the air in the upper atmosphere, since it is apparent that the whole path from a spectroscope at 35,000 ft. to the limits of the atmosphere contains about as much water as 2 m. of ground-level air. The spectroscope used to obtain these spectra contained a very small fluorite prism and used as a detector a bolometer with a strip of thermistor as a sensitive element. This detector, although a good deal less sensitive than a thermocouple, had to be employed on account of the vibration met with in a Mosquito. An electronically controlled

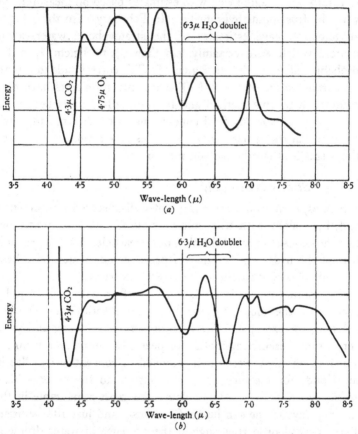

Fig. 25. (a) The 6·3μ band in the solar spectrum recorded at 35,000 ft. on 31 January 1952, 1520 G.M.T. Solar zenith angle, 79° 41′. (b) A spectrum of a Nernst filament showing the 6·3μ band after radiation has passed through about 2 m. of laboratory air.

heliostat, with a response time of about 0·1 sec., was installed, but even this speed of response was insufficient to overcome completely the most rapid aircraft oscillations.

The difficulties in the way of interpretation of solar spectra will be apparent when allied problems are considered in Chapter VI.

So far it has only been possible to make a rough interpretation based upon the shape of the band, which gives the total water vapour in the zenith within a factor 2. The results of eight flights to 20,000 ft. and above have been evaluated in this way, leading to results which agree well with estimates based on measurements with the frost-point hygrometer. In February 1949 three flights at 30,000 ft. gave estimates of the stratospheric water-vapour content which were certainly less than 7×10^{-3} g.cm.$^{-2}$, with a probable value of 1.4×10^{-3} g.cm.$^{-2}$. This corresponds to a mean mass ratio to air of 7×10^{-6}. Too few measurements have been made both by this method and by the method described in the last paragraph for any final conclusions to be drawn, but it will be seen that the spectroscopic estimates give a value which is six times smaller than the balloon measurements.

3.25. *Stratospheric clouds*

Stratospheric clouds are a rare but well-observed phenomenon, and, by analogy with tropospheric clouds, it is usually assumed that they consist of water droplets or ice particles. This is probably true of the iridescent 'mother-of-pearl' clouds which have been carefully observed by Störmer (1948) in Norway.

The mother-of-pearl clouds must be of great beauty, to judge even from the most objective scientific descriptions. Quoting Störmer, for example, 'The colours were very strong and pure, very closely resembling mother-of-pearl. As can be seen from my old drawings they were often arranged in concentric lamellae in the cloud; for instance, from the outside to the centre—blue, green, red and lilac.' The colours are most remarkable in the glancing rays of the sun just before sunset and just after sunrise. These optical properties suggest the presence of water droplets, and one cloud, observed against the moon, exhibited a typical droplet corona 15–18° in diameter. Recent work upon the super-cooling of water has shown that there is nothing unreasonable in the existence of water in the clean air of the stratosphere.

The clouds are between 2 and 3 km. thick, and height measure-ments have been made upon nine of them between 1926 and 1946; the results of these measurements are shown in Table VIII.

The total range of heights (22·0–28·5 km.) is remarkably small, being only about twice the average thickness of the clouds.

The meteorological conditions required for the formation of mother-of-pearl clouds are not yet fully understood (Dietrichs, 1949). They are believed to be 'orographic clouds' resulting from the lifting of air in standing waves caused by wind blowing over the Norwegian mountains. It is known that under favourable conditions it is possible for such waves to have very large amplitudes in the stratosphere. No exact analysis has yet been made of the particular wave patterns involved or of the condensation mechanism, and consequently the observations have not added to our knowledge of stratospheric water-vapour concentrations.

'Noctilucent' clouds are, as their name suggests, visible only at night in the glancing rays of the sun. They are not so brilliantly coloured as the mother-of-pearl clouds (Vestine, 1934), being

TABLE VIII. *Measured heights of mother-of-pearl clouds*

Date	Dec. 1926	Jan. 1929	Feb. 1930	Jan. 1932	Feb. 1932	Feb. 1932	Feb. 1934	Jan. 1944	Feb. 1946
Height (km.)	27·7	24·1	23·0	27·4	24·8	23·2	24·7	28·5	22·0

golden or reddish brown near the horizon, white or bluish white at intermediary altitudes, and blue-grey near the zenith. They may be very luminous, causing visible shadows from objects on the ground, and their intensity has been compared to that of the moon on the eastern horizon when the sun is setting in the west. The light from the clouds is polarized, and since the spectrum resembles that of the sun it is likely that it is scattered sunlight. All the height measurements which have been made give results very close to 80 km. The form of these clouds, which are extremely rare, is described in more detail in Chapter V, since they are mainly of interest as indicators of atmospheric motions at high levels.

The composition of the clouds is still a matter of debate, and both ice crystals and volcanic dust have been proposed. In favour of the former is the fact that the clouds are in a state of continual dissolution and reformation which is reminiscent of the condensation-evaporation process which takes place in cumulus clouds, for example. On the other hand, we have already mentioned in

§2.61 how an ice-crystal hypothesis demands either higher humidities or lower temperatures at the stratopause than present evidence suggests. The volcanic-dust theory fares no better, however, since no correlation can be found between the frequency of cloud occurrence and volcanic activity.

3.3. Carbon dioxide

There has only been one precise measurement of stratospheric carbon dioxide concentration, that made from samples collected on the Explorer II (1938) flight at 21·5 km., which gave a volume proportion of 0·029 ± 0·002 %. For information as to whether this is representative of the whole stratosphere, it is necessary to depend upon circumstantial evidence, which is, however, rather definite in its indications.

Carbon dioxide is both created and destroyed at ground level by animal life, plant life and bacteria. It is transferred upwards by mixing processes and is presumably destroyed in the upper atmosphere by photochemical decomposition, although this sink is probably well above the stratopause (Bates and Witherspoon, 1952). Diffusive separation, chemical and photochemical reactions and condensation all tend to prevent uniform mixing, but if over a great range of heights none of these factors is of importance it may be assumed that the gas concerned is uniformly mixed. Condensation is certainly not important in the case of carbon dioxide, and as far as can be judged from theoretical arguments, chemical and photochemical destruction are very slow below the stratopause, and would not be expected greatly to modify the distribution. Below the 'diffusion level' therefore (see §3.42) we may tentatively assume that the samples collected on the Explorer II ascent are representative of stratospheric air in general.

The concentration of carbon dioxide near the ground is of course variable, and it must be decided by observation whether this variability is of more than local importance. In cities and very near the soil the concentration is high, but away from the immediate vicinity of industrial areas, and only a few feet away from the soil it departs very little from that found in the Explorer II samples. Glückauf (1944), working at Kew, found the maximum range of concentrations to be 0·031–0·035 % by volume, while

Carpenter (1939), at Baltimore, U.S.A., found a range of 0·028–
0·033 % from 790 analyses. Samples collected between 4 and
10 km. by Dines (1936), whose apparatus is described below,
contained on the average 0·025 % of carbon dioxide and did not
vary significantly from the mean.

The sampling flask, contained in a light bamboo frame to protect
it from damage on landing, is carried into the stratosphere by a
small ·hydrogen balloon. When the balloon bursts, the change in
tension on the suspension thread closes a switch which connects

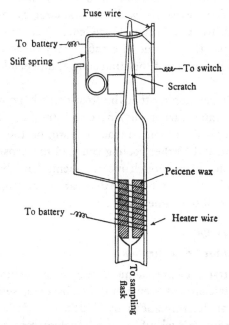

Fig. 26. Dines's sampling arrangement (schematic).
(After Dines, 1936.)

a battery across the fuse wire shown in fig. 26. The fuse wire, upon
melting, releases a stiff spring which strikes the glass tube sealing
the sampling flask, causing it to break off at a scratch. Stratospheric air then streams into the evacuated flask through a small
hole in a block of Peicene wax. Meanwhile a heater, which is
connected as the spring is released, softens the wax, and after a
short time the small hole is completely sealed.

The whole procedure is designed to avoid any possible contamination of the sample and, in particular, to avoid the use of tap grease. Despite these precautions the first experiments gave very erratic results for the carbon dioxide concentration, but eventually it was found to be possible to obtain consistent results. The analysis of carbon dioxide was performed by freezing out the carbon dioxide and water vapour together, and then measuring the vapour pressure of the mixture in an ice-bath. Twelve results from altitudes between 4 and 10 km. gave a mean of 0·025 ± 0·001 %, the measured extremes being 0·030 and 0·024 %. The agreement of these values with the Explorer II measurement, together with the circumstantial evidence which has been mentioned, suggests that a value of 0·026 % may be accepted for the carbon dioxide content of the stratosphere, independent of season or location.

It is worth mentioning that the frost-point hygrometer can be used to determine carbon dioxide concentration. By means of very rapid cooling, ice can be made to form on the thimble in a glassy deposit, and further cooling can lead to deposition of solid carbon dioxide. Some rough measurements have been made in this way (Dobson and others, 1946) near to the tropopause, and they agreed with the value given above.

3.4. The rare gases

3.41. *Measured concentrations*

The importance of precise measurements of helium, argon and neon concentrations has been established by a long series of careful microchemical determinations by Glückauf and Paneth (1945). Their technique (Glückauf, 1945 b) involves repeated fractional adsorption of air on to nut charcoal cooled in liquid nitrogen, which takes up ten times as much neon as helium, leading to a fairly efficient separation of these two gases.

The air to be analysed is dried and deprived of carbon dioxide chemically. Hydrogen is removed by means of a heated palladized platinum spiral. No hydrogen was ever found in samples collected at ground level, although small quantities were detected in balloon samples, which was taken to indicate contamination by the balloon itself. Oxygen is removed by heating a copper spiral, and the resulting change in volume can be measured. The remainder, a mixture of nitrogen and the rare gases, is then repeatedly fractionated through charcoal tubes immersed in liquid nitrogen in a manner designed to remove the nitrogen and argon and to procure effective separation of the helium and neon. The helium and the neon can in fact be brought to a very high state of purity before they are measured on a calibrated Pirani gauge; a small correction can be made for the impurity remaining in each sample. In its final form the apparatus has been proved capable of measuring neon and helium to within 0·2 % from a sample of air of volume 0·1 cm.³ at S.T.P.

With this apparatus the helium concentration at ground level has been studied. Samples collected from forty-one places distributed fairly evenly over the two hemispheres gave variations never greater than the estimated error of measurement, leading to a best value of $5·239 \pm 0·002 \times 10^{-4}$ % by volume. To within the errors of measurement, the ground-level concentration of helium is therefore a geophysical constant, and since it is a very light gas, the slightest tendency towards diffusive separation in the atmosphere should be very easy to detect. The same may be said of neon, whose ground-level concentration is $1·821 \pm 0·004 \times 10^{-3}$ % by volume, although it will not be such a sensitive indicator since its density is nearer to that of air.

Using the Dines technique for collecting samples, a series of measurements has been made both in England and the U.S.A. up to altitudes of 25 km. The results displayed in Table IX give the percentage increase in the ratio $He/(N_2 + A)$ over its mean ground-level value.

The general picture presented by these figures is that there is an irregular tendency towards diffusive separation which seems to increase with height. The surpluses are, however, so small and so irregular that in 1945 Glückauf and Paneth gave as their opinion that 'as regards the helium content of the stratosphere up to 25 km., the only conclusion which can be safely drawn is that it is very nearly the same as on the ground. The small and varying amounts of surplus helium found in air samples from between 20 and 25 km. height are not likely to be the result of diffusive separation at these heights in our latitudes.' They concluded that the question could only be elucidated when samples could be collected far higher than 25 km.

In the last year or two it has been possible to collect samples of stratospheric air up to 72 km. from rockets, and the most recent and reliable results (Chackett, Paneth, Reasbeck and Wiborg,

TABLE IX. *Measured helium surpluses in the stratosphere*

Height (km.)	Surplus (%)	Probable error (%)
16·5	0·5	0·5
18·0	0·35	0·1
18·5	0·7	0·3
19·0	0·55	0·15
21·0	6·9	0·7
22·0	4·1	0·2
22·0	1·95	0·15
22·5	5·1	0·6
22·5	1·9	0·3
23·5	4·0	0·3
23·5	0·3	0·15
25·0	2·1	0·3

TABLE X. *Rocket results*

Sample	Height (km.)	Ratio to the ground-level value of the proportions			
		$O_2:N_2$	$He:N_2$	$Ne:N_2$	$A:N_2$
16A	49·6–53·6	0·61	0·96	0·98	1·00
25D	50·4–53·3	0·60	0·98	1·00	1·00
15B	53·6–57·7	0·12	1·00	1·00	1·01
19D	54·7–58·3	0·00	1·01	(1·45)	1·01
3B	55·4–65·5	<0·02	0·98	0·99	0·99
1B	61·1–72·0	<0·02	0·93	1·00	1·01
B6	64·3–67·0	0·07	1·44	1·08	0·93
B8	67·0–69·6	0·25	2·02	1·18	0·89
B9	69·6–71·8	0·03	1·49 (3·44)	1·07 (1·35)	0·95 (0·73)

1951) show great changes in the concentrations of helium, neon and argon with respect to nitrogen above 60 km. The results fall into two groups according to the sampling technique employed, the first six and the last three in Table X. For the first six a glass sampling flask was used and heat was employed to make the seal. These are both considered to be bad features, and, in addition, there was some doubt as to whether these six samples were not collected by mistake from the ground-level air carried in the body of the rocket. It will be seen that apart from the increase of neon concentration in sample 19 D, which is considered to be in error, there is no sign at all of any change in the rare-gas concentrations from ground-level values. The decrease in measured oxygen concentrations, as has been mentioned, is only indicative of reactions in the sampling flask.

The last three sets of results, which come from one Aerobee flight, used the following sampling technique. An evacuated steel flask (which is preferable to a glass flask for helium determinations) was sealed with a closed copper tube. The flask was situated under the nose cone of the rocket, and this was jettisoned before sampling took place in order that the flask might be freely exposed. At the required height a steel knife severed the copper tube, and after 5 sec. it was squeezed shut by an electric clamp. The process was then repeated upon further flasks. All the seals were found to be effective except that of B 9, which had leaked slightly. Assuming that tropospheric air had entered it was possible to correct these results, and the corrected values are shown in brackets beside the measured values in Table X.

After making allowance for leakage in the B 9 sample, the last three sets of results give a very consistent picture of diffusive separation between 60 and 70 km. The light gases increase in concentration with height while the heavy gas decreases, leaving little doubt that the observations are reliable.

3.42. The diffusion level

The opposing tendencies of gravitation on the one hand, requiring an equilibrium with each gas distributed according to its own molecular mass, and mixing on the other hand, requiring that all gases are in constant relative proportions with the mean density determined by the mean molecular mass, have been the subject of a number of mathematical treatments. Exact results cannot be obtained, since the influence of steady mixing has not yet been put into useful mathematical form, and even if it were it would still be a long step to the erratic and variable conditions which exist in the stratosphere (see Chapter V). Some writers have, however, tackled the comparatively simple problem of the time required for the atmosphere to change from mixed to gravitational equilibrium, assuming that, starting from the former, mixing were suddenly to cease. Calculations show that the time taken depends upon the mean free path of the molecules to such an extent that it may be a matter of centuries at 20 km. (Brewer, 1949), of days from 100 to 150 km. (Maris, 1928) and of hours or minutes in the ionosphere (Mitra, 1948, p. 11). The calculated values depend slightly upon the particular gas under consideration and very greatly upon the assumed density at the level concerned; the reader is therefore advised to consult the original papers for more precise details. From these considerations it is possible to estimate the level below which the atmosphere is mixed, and above which it is in gravitational equilibrium (the diffusion level). The estimate depends upon the possible length of time for which the atmosphere might be undisturbed at various levels, but there is general agreement that above 400 km. the diffusion time is so very short that the atmosphere must be in gravitational equilibrium. Spitzer (1949) considers that the diffusion level must be above 100 km. for certain, and that the changes of composition are slight below 300 km. Mitra (1948) takes 10 hr. to be a possible estimate of the time available for diffusion, and concludes that the diffusion level must be near to 250 km. Maris (1928) took 5 days to be the critical time and found the diffusion levels for all atmospheric constituents to lie between 100 and 160 km.

The experimental results given in the last section have now to be considered in the light of this theoretical work. The remark by Glückauf and Paneth about the helium surpluses in the lower stratosphere, which we have quoted above, is now understandable, since the results described in Chapter V make it impossible to consider decades or centuries of quietude at these levels. If the helium results are correct (and this has not been seriously questioned), then they may indicate the erratic transfer of parcels of air from above the diffusion level, where their helium content is enriched in a matter of minutes, to the lower stratosphere, where they maintain their surplus until broken up by air motions. Such a picture, although of hopeless complication from the mathematical standpoint, is not at variance with other meteorological evidence. For example, Dobson and others (1946) have observed very dry layers of air persisting for days in a wet environment in the lower troposphere, which can only indicate that the air has been carried down from the upper troposphere or the lower stratosphere, and has remained unmixed.

The rocket results are even more difficult to explain. Theoretical workers are agreed in placing the diffusion level above 100 km., and this is not consistent with the changes of concentration observed at 70 km. With one set of observations, no matter how consistent internally, it is dangerous to assert that the theoretical work is seriously in error, and we must await more results and more precise theory before it is possible to decide between them. A decision is, however, of great importance, both from the point of view of measurements of temperature in the upper stratosphere, and from the point of view of the theory of the ionosphere.

3.5. Other minor constituents

Nitrous oxide and methane are of little importance in thermal studies of the upper atmosphere. Nevertheless, they are established atmospheric constituents, and information upon their vertical distribution is highly desirable.

3.51. *Nitrous oxide*

The solar spectroscopic technique of water-vapour determination which has been described in §3.24 can be equally well applied to the measurement of any gas with a strong absorption band. The strongest band of nitrous oxide is a doublet at $7 \cdot 8 \mu$, which, in a solar spectrum taken at sea level, is strongly overlaid by the wing of the $6 \cdot 3 \mu$ water-vapour band.

Fig. 27(a) shows a solar spectrum taken at sea level, with some evidence of a doublet centred at $7 \cdot 8 \mu$. This doublet might, however, occur without the presence of nitrous oxide, since groups of water-vapour lines in this region combine to give a somewhat similar appearance. At 5000 ft. a more convincing doublet is to be seen, while at 10,000 ft. the nitrous oxide band stands out very clearly, and its shape, when compared with laboratory spectra, shows that there is little interference by other absorptions. Some interference is to be expected since methane is known to have an absorption band at $7·7\mu$, but the agreement between the shapes of the solar and laboratory spectra indicates that no large error will be involved if, when analysing the results, the methane absorption is neglected.

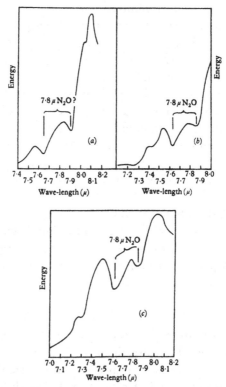

Fig. 27. The solar spectrum near $7·8\mu$ at various altitudes. (a) Sea level, solar zenith angle = $34°\ 16'$; (b) 5000 ft., solar zenith angle = $28°\ 38'$; (c) 10,000 ft., solar zenith angle = $42°\ 50'$.

The data for the interpretation of this band were obtained from a series of laboratory experiments (Goody and Wormell, 1951). From these it was possible to construct curves showing the relation between the band area and the product of the concentration and the secant of the zenith angle for various altitudes. These curves were based upon the assumption that the mixing ratio does not vary with height in the atmosphere, and it was realized that their use at various levels would prove whether or not the assumption is self-consistent. In practice no inconsistency was found, although the difficulties of observation were such that the test was not as discriminating as could be desired. Six results between 10,000 and 25,000 ft. gave a mean value for the volume concentration of $2.7 \pm 0.8 \times 10^{-5}\%$, where the limits give the spread of results, which may be due to error or partly due to real variations. A similar analysis made upon seven of Adel's spectra, taken in October and November 1938 and January, February, March, April, May and June 1939, gave the result $4.0 \pm 0.5 \times 10^{-5}\%$, with no obvious correlation between mixing ratio and the month of the year. To these results may be added those of Slobad and Krogh (1950), who measured the ground-level concentration at Dallas, Texas, using a mass-spectroscopic technique, and obtained values ranging from 2.5 to $6.7 \times 10^{-5}\%$ with a mean value of $5.0 \times 10^{-5}\%$.

While it might be possible to discern a slight tendency for the mixing ratio of nitrous oxide to decrease with height, the spread of results is consistent with an average value of $3.5 \times 10^{-5}\%$ through most of the troposphere.

3.52. *Methane*

It is theoretically possible to determine the vertical distribution of a gas by means of ground-level spectroscopic observations on the setting sun, for if the zenith angle (z) is greater than $85°$ (sec $z > 10$), then the intensity of an absorption line begins to depend upon the actual distribution of the gas in the atmosphere. Goldberg (1951) has made low-sun observations upon three lines in the $2\nu_3$ band of methane centred at 1.666μ. Using a very high resolution infra-red spectrometer (Goldberg, 1950), he could isolate the 0–1, the 5–6 and the 10–11 rotation lines and measure their areas as a function of sec z. Calculations were then made of the relations which would hold, provided that the lines had the Lorentz shape, for various vertical distributions of the gas. The only two really discriminating observations made were from Mount Wilson (1742 m.) on the 5–6 line for values of sec $z = 25$ and 60. These two observations, taken under difficult conditions, showed that the methane is not concentrated at high levels in the atmosphere, although they did not point with any precision to a mixed distribution.

Assuming a mixed distribution, however, it was possible to calculate the amount of methane in a vertical column above the level of observation and the line width at this level. Measurements made at Mount Wilson and Lake Angelus (296 m.) gave concordant results which were not incompatible with a mixed distribution and with the assumption of a Lorentz line shape. The half-line width at Lake Angelus was found to be $0 \cdot 18$ cm.$^{-1}$, and the number of methane molecules per cm.^{2} above the same place was found to be $3 \cdot 2 \times 10^{19}$, corresponding to a volume proportion of $1 \cdot 6 \times 10^{-4}$ % for a mixed atmosphere.

In contrast to this result, mass spectrographic analyses of ground-level air have given values of $5-10 \times 10^{-6}$ % for the methane concentration (Slobad and Krogh, 1950). Bates and Witherspoon (1952) have shown that such low concentrations at ground level are incompatible with theoretical ideas which indicate that the gas owes its existence in the atmosphere to seepage from mines and anaerobic decay of vegetation. Goldberg's result therefore appears more likely to be correct.

CHAPTER IV

OZONE

4.1. Introduction

Absorption by atmospheric ozone produces a cut-off for solar and stellar spectra at about 3000 Å., and this cut-off has been known ever since such spectra have been recorded. It was Cornu who first suggested that it must be caused by atmospheric absorption, and he believed that it was possible to obtain spectra extending farther into the ultra-violet by observing from high-level observatories. This was, however, never confirmed by later workers, who therefore came to the conclusion that the absorbing constituent did not reside in the lower levels of the atmosphere. In 1880 Hartley, who spent much of his life working on ozone problems, discovered the broad absorption band at 2600 Å. which bears his name and made the suggestion that this band was responsible for the spectral cut-off at 3000 Å. This speculation was confirmed finally by the work of Fowler and Strutt (1917), who showed that some weak bands discovered by Huggins in the spectrum of Sirius corresponded exactly to weak bands in the ozone spectrum.

Various chemical methods have been employed to measure the ozone concentration at ground level, but no great confidence was placed in the earlier results (see §4.23). The first reliable measurement of ground-level concentration was by Strutt (1918), who recorded spectra of a hydrogen lamp through four miles of air. Owing to some difficulties with atmospheric scattering Strutt could not make an exact measurement, but he could show that the ground-level concentration was far less than the assumption of a mixed atmosphere would suggest. From this it followed that the ozone existed somewhere in the upper atmosphere.

Cabannes and Dufay (1925) were the first to measure the mean height of the atmospheric ozone, which they did by means of low-sun spectral observations. They arrived at a value of 50 km., which later measurements show to be too great by a factor 2.

It only remained now to measure the actual vertical distribution, and this was done simultaneously by the Regeners (1934) and Götz, Meetham and Dobson (1934). The former sent a small solar spectroscope up on a balloon to 31 km., while the latter obtained their results from measurements on the scattered light from the zenith sky. The two sets of measurements agreed remarkably well.

It was soon recognized that the variation in the amount of ozone was a factor of some importance, since it should be connected with events in the upper atmosphere. Fabry and Buisson (1921) pioneered the technique of precise measurement, and this was followed up by Dobson and Harrison (1926) and Dobson (1931). Dobson and his collaborators have since been responsible for much of the systematic collection of data which has taken place.

To complete this brief historical survey it should be mentioned that Chapman (1930) was the first to devise a theoretical treatment of the formation of atmospheric ozone. His work has been improved and greatly extended by many subsequent workers.

Since it is largely through its absorption spectrum that ozone is of such importance, it is necessary to give a brief sketch of its characteristics. Gaseous ozone absorbs in bands in the ultra-violet, visible and infra-red regions. The main feature of the ultra-violet spectrum is the broad system of Hartley bands, which take the form of indistinct peaks upon a strong continuous background from 1800 to 3400 Å., with the band centre at 2600 Å. On the long-wave wing of the Hartley bands is another band system, of a different nature, called the Huggins bands. These are much weaker but much sharper than the Hartley bands, and they lie between 3200 and 3600 Å., with their heads towards the red

Absorption coefficients in the Hartley and Huggins bands have been measured with a care commensurate with their importance to atmospheric physics. Craig (1950) has considered all the laboratory measurements made and has given the curve shown in fig. 28(a) as his opinion of the best available information. This curve is mainly based upon the results of Ny and Choong (1932, 1933), although all structure has been smoothed out, thus obscuring the distinction between the two band systems. A more detailed representation of the Huggins bands is given in fig. 41.

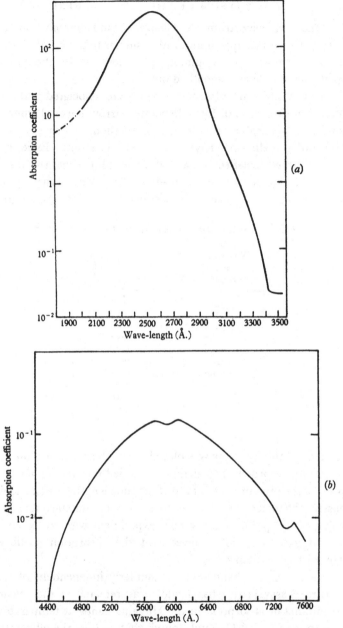

Fig. 28. (a) The Hartley and Huggins bands. (b) The Chappuis bands. (After Craig, 1950.) The absorption coefficients aᵣ to the base *e*, with unit ozone amount equal to 1 cm. at s.t.p.

In the visible spectrum, the Chappuis bands stretch from 4400 to 7400 Å., with an appearance rather similar to the Hartley bands. Fig. 28(b) shows Craig's opinion of the best data on these bands, again with the details smoothed out.

All the visible and ultra-violet bands are associated with electronic transitions, and, for sufficiently narrow spectral intervals, Lambert's absorption law is obeyed (without which the words 'absorption coefficient' have no obvious meaning). Experiment has established that Beer's law is also obeyed, i.e. that the absorption coefficients are not pressure-dependent. Vassy (1935) passed light through two absorption tubes, one forty times the length of

TABLE XI. *Temperature dependence of the Huggins bands*

Wave-length of absorption minimum (Å.)	$\dfrac{k(-80°\ \mathrm{C.})}{k(+20°\ \mathrm{C.})}$
3359	0·40
3327	0·44
3295	0·46
3268	0·50
3239	0·51
3213	0·67
3190	0·72
3168	0·77
3151	0·85
3130	0·92
3110	0·99

the other. Initially ozone was placed in the shorter tube alone and a spectrum recorded. It was then expanded into the longer tube, lowering the pressure in the ratio of the tube lengths whilst keeping constant the total ozone. The spectrum was found to be unchanged. Strong (1941) performed a similar experiment with an expansion ratio of 750 to 1, and showed that the absorption coefficient changed by less than 5 %.

These electronic bands are not similarly independent of temperature. Vassy (1935) has made measurements at the maxima and minima of absorption in the Huggins bands at temperatures of −80 and +20° C., and found that, while the absorption maxima were independent of temperature, the minima were strongly dependent, as shown in Table XI. The quantity given in

Table XI is the ratio of the absorption coefficient at −80° C. to that at +20° C.

Absorption in the Chappuis band is also temperature-dependent. Vassy (1937) has shown that the maximum variation takes place at the band centre, where a change of temperature from −80 to +20° C. will change the absorption coefficient by 20%.

In the infra-red region of the spectrum, ozone has three intense vibration-rotation bands and a number of weaker overtone bands. Fig. 29 shows a spectrum recorded by Hettner, Pohlmann and Schuhmacher (1935).

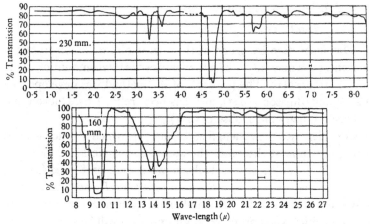

Fig. 29. The infra-red absorption spectrum of gaseous ozone. (After Hettner and others 1935.) The figures in the diagram refer to the pressure of pure ozone in a 30 cm. tube.

Prominent bands can be seen at 4·75, 9·6 and 14·1 μ, of which the 9·6 μ band is the strongest. According to the latest interpretations of the spectrum (Wilson and Badger, 1948) there is another strong band at 9·0 μ, which may be seen in Hettner, Pohlmann and Schuhmacher's spectrum to be nearly as strong as the 14·1 μ band but which has, until recently, been overlooked, since it is so heavily overlaid by the 9·6 μ band.

At slightly higher resolution all the bands in fig. 29 would appear as doublets, and at much higher resolution each would show a complicated line structure. For high enough resolution Lambert's law would be obeyed, although such resolution cannot be obtained by the best of modern spectrometers. Beer's law is, however, never obeyed, since the line structure is pressure-dependent. Moreover, both the strengths and widths of the individual lines depend upon temperature. These are problems common to all vibration-rotation bands, and their treatment is fundamental to the atmospheric problem. For this reason they will receive more detailed consideration in Chapter VI, and they are only mentioned here to show that the problem of absorption in the infra-red bands is of immensely greater complication than that in the visible and ultra-violet ozone bands.

4.2. Methods of measurement

4.21. Total amount of ozone

The absorption by ozone of the ultra-violet radiation from the sun and stars may be utilized to measure the total amount of ozone in the atmosphere. The principle of the method was developed by Fabry and Buisson (1921), and has been refined by Dobson and Harrison (1926) and Dobson (1931).

Consider two wave-lengths, at one of which ozone absorbs strongly, while at the other the absorption is weak. (The wave-lengths used originally by Dobson were 3110 and 3290 Å.) Let the solar intensities outside the atmosphere be I_0' and I_0 at the two wave-lengths. According to Lambert's exponential absorption law the intensities when observed at ground level will be given by

$$\log I' = \log I_0' - (\beta' + \delta') \sec z - \alpha' x \sec \chi_h,$$
$$\log I = \log I_0 - (\beta + \delta) \sec z - \alpha x \sec \chi_h,$$

where β, β' = attenuation coefficients for molecular scattering,

δ, δ' = attenuation coefficients for scattering by dust, water droplets, etc.,

α, α' = ozone absorption coefficients,

x = total amount of ozone per cm.^{2} in a vertical column of the atmosphere,

z = solar zenith angle at the ground,

χ_h = solar zenith angle near the centre of gravity of the ozone.

The two zenith angles have to be used since the scattering occurs mainly near to the ground while the ozone absorption occurs mainly at 25 km., and for a curved atmosphere zenith angle is a function of height. The relation between them is simply

$$\sin \chi_h = \frac{\sin z}{1 + h/a},$$

where a = the radius of the earth, and h = the mean height of the ozone.

All the quantities in the above two equations may be calculated or measured, except x and the particle attenuation coefficients. Since the particles concerned are presumed to be much greater than the wave-length of the radiation involved, it has usually been assumed that the scattering is 'white', and therefore that $\delta = \delta'$. This fundamental assumption has recently been questioned by Ramanathan and Karandikar (1949), who believe that it is not valid under hazy conditions. Under other conditions, however, the two equations may be subtracted to given one relation not involving δ:

$$\log \frac{I'}{I} = \log \frac{I_0'}{I_0} - (\beta' - \beta) \sec z - (\alpha' - \alpha) x \sec \chi_h.$$

From laboratory measurements and theoretical calculation Dobson used the values $\alpha' - \alpha = 1 \cdot 153$ and $\beta' - \beta = -0 \cdot 08$ for the two wave-lengths in question. The value used by Dobson for $\alpha' - \alpha$ is based upon work by Fabry and Buisson (1913) and does not agree with the more recent work of Ny and Choong (1932). Götz (1944) has pointed out that the use of the older coefficients may lead to values of x which are 12 % too high. It is very likely that these coefficients will be further modified as more precise experiments are made, but the matter is not serious, since the interest in the ozone problem lies more in the determination of small changes in the ozone amount rather than its precise magnitude.

The values of $\log I_0'/I_0$ (called the 'extra-terrestrial constant') and h are found by experiment. It is assumed that upon a perfectly settled day the ozone amount remains constant, and measurements are taken over a very wide range of zenith angles. A value of h is then found which leads to a linear relation between

$$\log I'/I + (\beta' - \beta) \sec z$$

and sec χ_h, and the intercept is taken to give the value of the extra-terrestrial constant. The value of h is not at all critical, and a mean value of about 25 km. can safely be used at all times.

The only other uncertainty in this method of measuring ozone amount is the possibility of variations of the extra-terrestrial constant. According to Watanabe (1943), its value for the wave-lengths used by him may vary sufficiently to cause errors of 13 % in the measurement of x. Dobson, however, does not consider this to be a serious source of error.

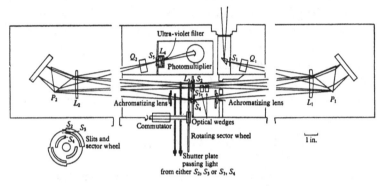

Fig. 30. Dobson's ozone spectrophotometer.
(After Normand and Kay, 1952.)

One of the standard instruments for measuring the ratio log I'/I is a two-prism spectrophotometer designed by Dobson; a dia-grammatic sketch of this instrument is shown in fig. 30.

Radiation entering the spectrophotometer passes through the entrance slit S_1, and is then collimated by the lens L_1, doubly dispersed by the prism P_1 and refocused by L_1 on to the three slits S_2, S_3 and S_4 (the wave-length passing S_4 is in the range 4355–4740 Å., and is used for purposes which will not be discussed here). Since the focal length of L_1 depends upon wave-length, the achromatizing lens is required to focus the three selected wave-lengths on to the plane of the slits. The lens L_3 and the small prism behind S_4 are required to ensure that all the radiation passing the three slits falls upon the lens L_2. The lens and prism system L_2P_2 recombines the three wave-lengths on to the slit S_5, behind which is the photomultiplier detector.

The main purpose of the instrument is to balance the intensities passing S_2 and S_3, both of which correspond to wave-lengths in the Huggins bands. For this purpose, a sector wheel (shown also at the bottom left of fig. 30) interrupts the two beams in turn at 22 c./sec., and the optical wedges in front of S_3 are adjusted until no alternating component can be detected by the photomultiplier; the position of the wedges then gives a direct estimate of the ratio of the two intensities. The commutator shown gives synchronous rectification of the amplified photomultiplier signal, while the plates Q_1 and Q_2 are used for temperature compensation, and also to vary slightly the chosen wave-lengths.

It will be seen from fig. 30 that a great deal of trouble has been taken by means of many diaphragms and the double dispersion to ensure the greatest degree of spectral purity, i.e. absence of scattered radiation. This is important because the intensity of radiation at the shorter wave-length may be minute compared with the total intensity of the longer wave-lengths which enter the instrument. As a result of the various precautions taken Dobson claims that his instrument is capable of measuring the ozone amount within 2%.

With modern photomultipliers it is possible to use this apparatus with unfocused solar rays scattered from a ground-quartz plate over the entrance slit even for measurements with the sun very near the horizon. With the plate removed observations can be made upon the scattered light from a clear sky or upon the light from the base of clouds. Measurements have also been made using the moon as a source (Normand and Kay, 1952), but for this purpose a focused image is necessary.

Methods have also been developed which make measurements possible when direct sunlight does not reach the instrument. Since nearly all scattering takes place low in the troposphere, except when the zenith angle is very large, light received at the ground has passed through the ozone layer before being scattered. Thus the correct value of the zenith angle is known, and the effect of the scattering is only to modify the value of the extra-terrestrial constant. This modification can be determined empirically, and ozone determinations can be made almost as accurately with overcast as with clear skies.

4.22. *The vertical distribution of ozone*

4.221. *Ground measurements.* It has already been mentioned that a value for the mean height of the ozone in the atmosphere may be found by measuring intensities at two wave-lengths for large solar zenith angles. Fig. 31 shows log I'/I plotted as a func-

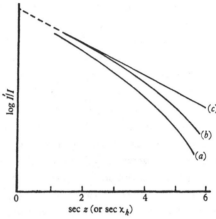

Fig. 31. Determination of the mean height of the atmospheric ozone. (After Götz and others, 1934.)

tion of sec z for large values of z (curve (a)). The relation is not linear and is not made so by adding values of $(\beta - \beta')$ sec z to the ordinate (curve (b)). A value of h is now sought which gives a linear relation between log I'/I and sec χ_h (curve (c)). This is not a very precise method of determining h, but with care it can lead to values near 27 km. which agree fairly well with the more discriminating methods which will now be described.

If a spectrophotometer is adjusted to receive scattered light from the zenith sky instead of direct sunlight, then, for reasons which have already been mentioned, curves like that shown in fig. 31(a) are still obtained for moderate zenith angles. For very large values of sec z, however, the value of log I/I reaches a minimum and subsequently increases. This effect, named the 'inversion effect' by Götz, is shown in fig. 32 from results taken at Delhi by Karandikar and Ramanathan (1949). Götz (1931a), besides discovering the inversion effect, realized that it could be used to find the vertical distribution of ozone.

The origin of the inversion effect is illustrated in fig. 33, which shows a stratified atmosphere with the ozone (total amount, x) concentrated in a narrow layer. Let the subscript 1 refer to a level below the ozone and the subscript 2 to a level above the ozone. The scattered light from the zenith sky received at ground level will vary with the number of molecules per cm.[3] (n) at the level concerned and with the attenuation caused by atmospheric absorp-

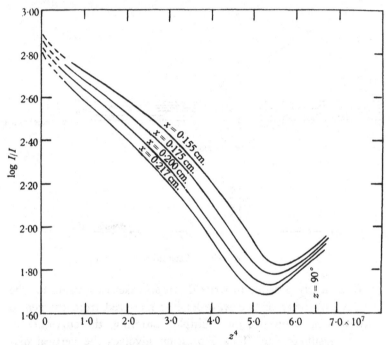

Fig. 32. The inversion effect measured at Delhi. (After Karandikar and Ramanathan, 1949.) x = total ozone amount; z = solar zenith angle.

tion. The light received from the lower level will therefore be proportional to $n_1 \exp(-\alpha x \sec z)$, while that from the upper level will be proportional to $n_2 \exp(-\alpha x)$. Now, although n decreases very rapidly with height, if $\sec z$ is sufficiently large more scattered light may be received from the upper level than from the lower. There is, in fact, a level of maximum scattering to the ground which, as $\sec z$ increases, varies from ground level to above the ozone layer. The spectrophotometer, however, measures the total ozone in the path, and therefore for small zenith angles it measures $x \sec z$, while for very large zenith angles it must measure only x. Thus, as the zenith angle increases, the measured ozone from observations on the zenith sky first of all increases but subsequently falls to its value at zero zenith angle. Since $-\log I'/I$ is, in effect, a measure of the total ozone in the path, this leads to the observed inversion effect.

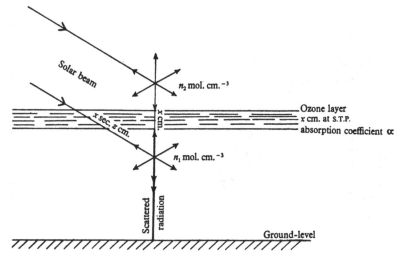

Fig. 33. The origin of the inversion effect.

It is clearly possible to write down an exact expression for the relative intensity of two wave-lengths scattered from the zenith sky, making allowance for multiple scattering, the curvature of the atmosphere, etc. This expression involves the vertical distribution of ozone in such a complicated way that it is not possible to calculate it rigorously from the observed inversion curves. Two simpler methods have been used. In the first the atmosphere is divided into five layers, viz. 0–5, 5–20, 20–35, 35–50 and above 50 km. The ozone contained in the lowest layer is assumed to be known from ground-level chemical analyses, while the highest layer is assumed to contain no ozone. Since the total ozone is known, this leaves only two parameters to be determined. The scattering law is greatly simplified by assuming that only single scattering from molecules (i.e. Rayleigh scattering) is important.

It is then possible to calculate the expected value of $\log I'/I$ at various zenith angles as a function of the two unknown concentrations. These can then be determined by a graphical method from observations at any two zenith angles.

An assumption of single scattering, however, is difficult to justify. For a clear sky, approximately 77 % of the brightness is due to single scattering, 12 % to double scattering and about 3 % to scattering of a higher order. The remaining 8 % is due to light which has been scattered or reflected from the ground. The inclusion of multiple scattering into the calculations is exceedingly difficult. While the singly scattered light from any level may have had to pass through the overlying ozone at a large zenith angle, a proportion of the multiply scattered light will have come through at nearly perpendicular incidence, and will have suffered less absorption. The ratio of singly to multiply scattered light received at ground level in the ozone absorption bands is therefore a complicated function of the ozone distribution and the zenith angle. Both of the simple methods of analysing the inversion effect which are described here have necessarily assumed single scattering, since the full theory of multiple scattering and absorption has not yet been worked out.

For the second method of analysis the atmosphere is divided into eight layers below 80 km., above which level there is assumed to be no ozone. It is found that the inversion curve for very large zenith angles is dependent mainly upon the ozone amount in the uppermost three layers, and this can therefore be independently determined. The total ozone is obtained from measurements at small zenith angles, and then trial inversion curves are constructed using the remaining four variables until a good fit is obtained to the observed results. Small errors in observation can lead to great changes in the optimum amounts in the middle layers.

Fig. 34 shows comparative calculations of the vertical distribution of ozone over Arosa by the two methods which have been described (Götz and others, 1934). The two methods lead to similar results, with the maximum concentration between 25 and 30 km. and with the centre of gravity between 20 and 30 km.

It is very difficult by such inexact methods to detect any fine structure in the vertical ozone distribution. Nevertheless, Götz

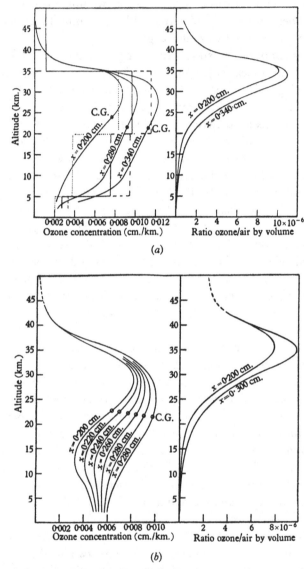

Fig. 34. The vertical distribution of ozone at Arosa from measurements of the inversion effect. (After Götz and others, 1934.) x = ozone amount; C.G. = centre of gravity of the ozone. (a) Using five layers; (b) using eight layers.

(1938, 1944) and Karandikar and Ramanathan (1949) both believe that their results indicate that for large ozone amounts there are two maxima of ozone concentration, one near 30 km. and one lower down and rather variable.

4.222. *Airborne apparatus.* The first direct measurements of the vertical distribution of ozone were made by E. and V. H. Regener (1934) at Stuttgart, at the same time as the first inversion effect experiments were made at Arosa. A small ultra-violet spectrometer with a quartz prism (aperture 25 mm. at $f/$ 12 with 0·01 mm. slits) was suspended in a protective gondola (see §1·41), pointing downwards at a gypsum scattering plate illuminated by direct sunlight. The weight of the apparatus was 2·7 kg., and two small hydrogen balloons could lift it to 31 km. Photographic records of the spectrum were taken at regular intervals up to the ceiling, each lasting about 10 min. Each successive spectrum was found to extend farther into the ultra-violet than those from lower levels, and at 29·3 km. the extension, as compared with ground-level spectra, was 200 Å.

The total ozone above the balloon can, with a few refinements not used in the earlier ascents (Regener, V. H., 1938), be measured to within 1 % of the total above ground level. Frequency calibration is possible *in situ* from the Fraunhofer lines recorded on the spectrum, and intensity calibration is made by the standard astronomical technique of crossing the slits with a number of filters which give known attenuations. The accuracy is sufficient to enable the concentration of ozone to be obtained from the slope of the total amount versus height curve.

Fig. 35 shows a number of recent measurements made by this method over New Mexico (Regener, V. H., 1951). These results differ a little from earlier measurements which have been made. Usually a simple maximum of concentration is found near 25 km. This maximum is normally not nearly so sharp as on the 25 February flight, although a very similar result was obtained on the only occasion when ozone was measured from a manned balloon (Explorer II, 1938). There is no reason to doubt any of these measurements, and it must therefore be concluded that the vertical distribution of ozone varies considerably from time to time.

Another point of interest about the 25 February flight is the fact that the ozone to air mixing ratio appears to be nearly constant above 20 km., suggesting considerable mixing, since, as will be seen in §4.4, equilibrium theory requires a rapid decrease of ozone with height above the maximum concentration.

The flight on 18 April gives some indication of a double maximum of ozone concentration. Such a distribution has also been suggested by the inversion-effect measurements.

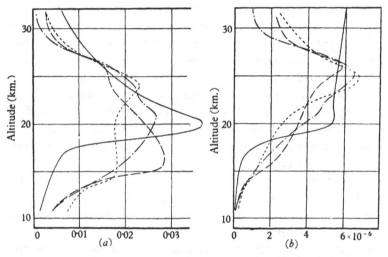

Fig. 35. The vertical distribution of ozone over New Mexico during 1950. (After Regener, 1951.) (a) Ozone concentration (cm./km.); (b) proportion of ozone to air molecules. —— 25 February; —·— 4 March; ——— 16 March; ···· 18 April.

A number of attempts have been made to develop an ozone radio sonde in order to make routine measurements. There are more difficulties in such work than are apparent, but nevertheless one instrument has been flown on a number of occasions (Coblenz and Stair, 1939, 1941). For the sake of cheapness a filter system is used. A photoelectric cell views a scattering plate illuminated by sunlight either through one of two filters, each capable of isolating a region in the Huggins bands, or through one of two metal grids used for standardization of the equipment. The two filters and the two grids are slowly rotated on a disk, and the resulting signal from the photoelectric cell is converted into a radio signal and thus transmitted to the ground.

Results obtained with this radio sonde have agreed fairly closely with the earlier spectroscopic measurements in showing a flat maximum of concentration near 25 km.

Solar ultra-violet spectra have also been obtained on rocket flights and the results analysed in terms of the ozone distribution. Spectra extending to 2000 Å. have been obtained from altitudes of 160 km. (Durand, 1949). The main difficulties in this work have been caused by the yawing and rotation of the rocket after fuel burn-out, since, owing to the short exposure times made necessary by the great speed of the rocket, it has not been possible to use diffuse reflecting surfaces as sources for the spectroscope. A number of devices have been used to obtain what amounts to a focused image of the sun on the slits, including rapid sun-followers, but the most ingenious and also the most successful apparatus has been the N.R.L. bead spectrometer.

In place of an entrance slit the N.R.L. spectrometer uses a 2 mm. bead of lithium fluoride which, acting as a very short-focus lens, gives a solar image 0·013 mm. in diameter. Owing to the very short focal length of this bead the position of the solar image will only change very slightly if the direction of the incident sunlight changes, and as a result the solar image can be used instead of an entrance slit. The dispersing system of the spectroscope is a concave grating with a centre of curvature of 40 cm. and with 15,000 lines ruled to the inch; the spectrum is recorded on a strip of 35 mm. film. In order to double the chance of radiation entering the spectrometer, two beads are used, one on each side of the rocket, and the two resulting spectra are focused on different parts of the film. The resolution of this spectrometer is a matter of chance, since it depends upon whether or not the blurring caused by rocket motions lies at right angles to the dispersed spectrum. Under favourable circumstances the resolution has been as good as 1 Å., which is a remarkable achievement.

Solar spectra obtained on three rocket flights have been interpreted in terms of the vertical distribution of ozone. Below 30 km. the ozone concentrations thus obtained do not differ in any marked way from those obtained from balloon ascents, and it is the

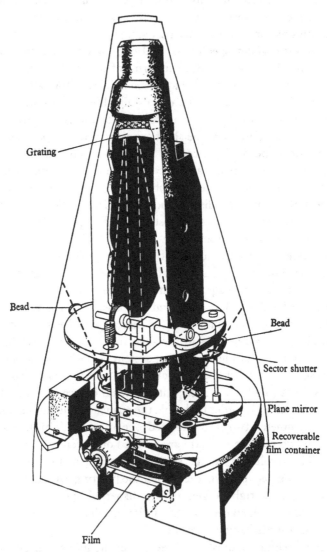

Grating

Bead

Bead

Sector shutter

Plane mirror

Recoverable
film container

Film

Fig. 36. The N.R.L. bead spectrometer. (After Durand, 1949.)

information which they give about the concentrations in the higher
levels which is of the greatest interest. On 19 June 1949 a rocket
firing was made at sunset with a solar altitude of only 1° in order
that the small ozone amounts above 30 km. should give the
greatest possible absorption. Measurable absorptions were observed on spectra taken as high as 70 km., and the ozone distribution could be determined up to this level. On this occasion

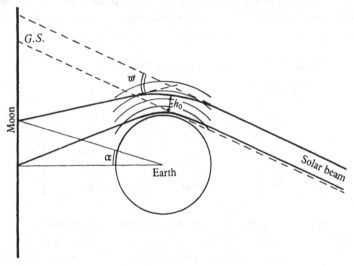

Fig. 37. Light paths during a lunar eclipse.

the maximum concentration was $1 \cdot 1 \times 10^{-2}$ cm./km. at an altitude
of 26 km. Above this level the concentration decreased exponentially with altitude, falling to $2 \cdot 5 \times 10^{-6}$ cm./km. at 70 km. This
result is of considerable significance when compared with the
ozone concentrations predicted from photochemical equilibrium
theory (see §4.42).

4.223. *Astronomical measurements.* During a lunar eclipse a
shadow of the earth's atmosphere is thrown on to the surface of
the moon, and it is possible by means of spectral observations to
draw conclusions about the vertical distribution of absorbing
constituents in the atmosphere.

Radiation passing through the earth's atmosphere is modified by scattering, by atmospheric absorption, and it is also refracted owing to the density gradient of the air. Scattering in the upper atmosphere is believed to be by molecules only, and therefore the scattering term can be calculated with some precision. The refraction effect is very helpful, since the atmosphere acts as a weak divergent lens and forms a magnified image of itself on the surface of the moon.

In fig. 37, h_0 is the distance of closest approach of a ray to the earth's surface and ϖ is the refraction angle. Co-ordinates on the moon's surface may be specified by means of the angle α. If

TABLE XII. *Deviation of rays during a lunar eclipse*

h_0 (km.)	ϖ	α
0	69·9′	−13′
2	55·5′	+ 1·6′
4	44·9′	+12·2′
10	25·1′	+32·2′
20	5·56′	+51·7′
30	1·16′	+56·2′
40	0·24′	+57·2′

there were no refraction the geometric shadow of the earth $(G.S.)$ would have a radius of 57·0′, while the shadow of the atmosphere up to 40 km. would have a radius of 57·5′. Table XII shows the effect of refraction by the atmosphere. The region up to 40 km. is magnified 120 times, while the important region from 10 to 40 km. is magnified 70 times (van de Hulst, 1949).

Spectrophotometric observations on the earth's shadow provide an interesting method of studying the ozone layer (Götz, 1931 b; Barbier, Chalonge and Vigroux, 1942). One of the rays from the sun will pass through a total amount of ozone equivalent to 11 cm. at S.T.P., and this makes it very difficult to use the Huggins bands since the absorption is too strong. Recently Paetzold (1952) has made measurements on the Chappuis bands in the visible spectrum with considerable success. Since each ray passes through all levels of the atmosphere above h_0, a differential method of analysis has to be employed which only gives good results where the ozone concentration decreases with height, i.e. above the ozone maximum. During one eclipse, however, Paetzold was able to place the maximum concentration at 21 km. and to show that the concentration fell to very small values above 50 km.

This is potentially one of the most effective methods available for the measurement of ozone at high levels and the only one free from the need for very expensive equipment. However, information is likely to come in rather slowly, since a lunar eclipse under perfect observational conditions is rare. There are also great complications which have not, so far, been mentioned (Link, 1933). The lunar surface is a very poor screen, being uneven and spotty, and the observations cannot therefore be of very high quality. Even more serious is the fact that the sun has an angular diameter of 32′, which almost ruins this promising experiment and makes the analysis extremely complicated.

4.23. *Chemical methods*

The main difficulty with the chemical measurement of ozone is that it is never specific, and other possible atmospheric gases can bring about reactions similar to those used to detect the ozone. Early measurements were therefore greatly distrusted upon this score, and it is only in recent years that it has been established that the oxidizing agent measured can be identified with ozone. The basic reaction used is

$$O_3 + 2KI + H_2O = 2KOH + O_2 + I_2,$$

followed by a titration of the free iodine with sodium thiosulphate, using starch or an electrolytic indicator. The latter depends for its action on the fact that free iodine can oxidize the polarizing layer of hydrogen on a platinum cathode, thus allowing a current to flow in an external circuit. The chemical difficulties in this method lie in the great care necessary with respect to impurities, on account of the very dilute solutions employed, and in the possibility that, in acid solution, the nascent oxygen may liberate more iodine from the potassium iodide. Careful handling of the re-agents and a close control of the hydrogen-ion concentration can reduce any errors involved to tolerable amounts.

The non-automatic method of Ehmert (1949*b*), now in routine use with the German Wetterdienst, is probably the most accurate of those which have been developed. The apparatus developed by

Glückauf, Heal, Martin and Paneth (1944) is probably the most accurate automatic instrument, but it is somewhat complicated, and we will therefore describe briefly a simpler version of it used by Bowen and Regener (1951), which is said to be capable of analyses within 5%.

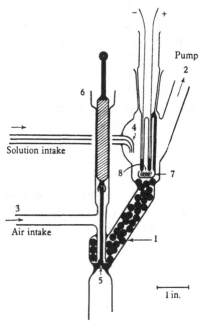

Fig. 38. Automatic chemical measurement of ozone.
(After Bowen and Regener, 1951.)

Fig. 38 shows a sketch of Bowen and Regener's apparatus. A metering pump, attached to (2), draws air at the rate of $1 \cdot 5$ l./min. into the intake (3) and through the reaction chamber (1). The reaction chamber is filled with glass beads to bring about the greatest possible surface area of contact between the air and the 8 cm.^{3} of mixed sodium thiosulphate and potassium iodide contained in it. The potassium iodide concentration needs only to be over a certain limit, but the sodium thiosulphate is made up with precision to a 5×10^{-6} normal solution. At first the liberated iodine is taken up by the sodium thiosulphate, but after this reaction has been completed a further 10% of iodine will depolarize the electrodes (8) contained in the electrolytic cup (7). The cathode takes the form of a helix of o·5 mm. platinum wire, and the anode lies along its axis. A potential difference of 22 mV. is used, and the external circuit is triggered by a current of 2×10^{-9} amp. The external servo-system, which is not shown in fig. 38, causes the used charge to flow out at the waste (5) and meters a new charge by means of a medical syringe (not shown) from a supply connected to the charge intake (4). With normal ozone concentrations at ground level the complete cycle of operations takes about 30 min., and this time is recorded as a measure of the ozone concentration.

Although such apparatus is capable of measurements in the stratosphere, only tropospheric measurements have been made up to the present time. Since these cannot be divorced from the problem of ozone in the stratosphere, we may mention some of the results obtained by Ehmert (1949a).

Fig. 39 shows the results of six sets of measurements made from an aircraft up to 9 km. during August 1942. It will eventually be of great interest to study such results in relation to atmospheric turbulence, since, as will be shown in §4.4, this is of great theoretical interest. However, the limited number of results shown indicates only the great variability of ozone concentration in the troposphere both with height and with time. It will be seen that there is a tendency for a maximum concentration to occur in the region of 4 km.

Fig. 40 illustrates two interesting correlations of the ground-level ozone concentration and other factors; (a) shows that during the great convective activity at midday the ground-level ozone increases, while (b) shows that the Föhn wind, which has passed over much higher levels, is rich in ozone but contains little water vapour. Both of these phenomena suggest that high ozone concentrations at ground level are to be associated with air being brought down from higher levels, either by mixing or by a vertical wind component.

4.24. *Temperature measurements*

4.241. *Measurements on the Huggins bands.* Measurements of the contrast between maxima and minima in the Huggins bands have been used by Barbier, Chalonge and Vassy (1935) to estimate the temperature of the atmosphere between 20 and 30 km., since, as has been described in §4.1, the contrast is temperature-dependent. Most of these measurements have been made at night upon A and B type stars, for the reason that these stars have an emission spectrum similar to that of a high-pressure laboratory

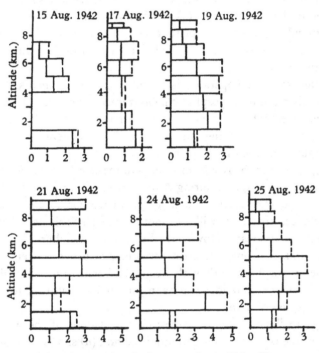

Fig. 39. Ozone concentrations in the troposphere. (After Ehmert, 1949a.)
—— ozone, cm./km. × 10³; ---- ozone to air ratio × 10⁸.

hydrogen lamp with a great proportion of the energy in the ultra-violet; this alleviates the problem of internal scattering in the spectroscope. Dufay (1936) has also made measurements on the scattered light from the zenith sky, and recently Kay (1951) has used the Dobson spectrometer for direct sun measurements.

The French workers have used a technique which gives an average figure for the nine Huggins bands detectable between 3140 and 3340 A. A number of photographic records are made in this region for a range of zenith angles. The extra-terrestrial constant is determined and allowance is made for Rayleigh scattering. According to Vassy (1935) the absorption maxima are not

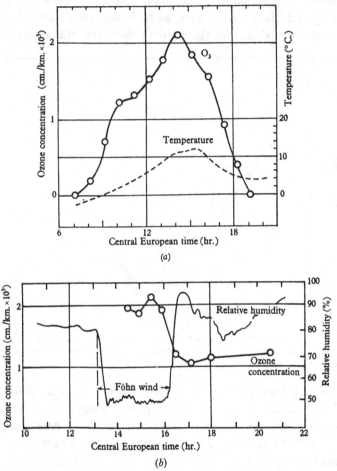

Fig. 40. (a) The diurnal variation of temperature and ozone concentration at Friederichshaven. (After E. Regener, 1949.) (b) Ozone concentration and humidity at ground level during the cessation of the Föhn wind. (After E. Regener, 1949.)

temperature-dependent, and therefore the laboratory results for these points are used to obtain the total amount of ozone. This result may then be used to obtain the absorption coefficients at

the absorption minima, which can then be compared with laboratory values.

One comparison of this nature is shown in fig. 41. The
agreement between the laboratory and atmospheric values at
the maxima is of no interest, since it has been forced by the
process of analysis. At the minima, however, the atmospheric

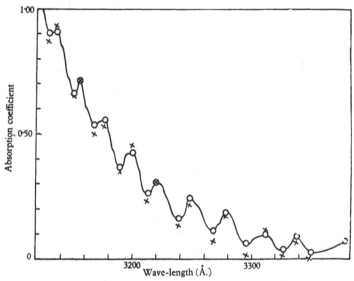

Fig. 41. Stratosphere temperature from the contrast in the Huggins bands.
(After Barbier, Chalonge and Vassy, 1935.) O from measurements at room
temperature; × from atmospheric transmission.

absorption coefficients are almost all low compared with laboratory
values, indicating that the stratospheric temperature is below
room temperature. By comparison with laboratory data obtained
at low temperatures, the stratospheric temperature may be
estimated.

Table XIII shows a series of measurements made by Barbier
and Chalonge (1939) in Switzerland and Lapland, grouped
according to the total amount of ozone. There is a marked correlation between temperature and ozone amount, and a similar effect
has been observed by the Vassys (1938) from zenith sky observations in Morocco. This correlation has been taken to indicate
that ozone changes are the result of discrete air masses moving

from one latitude to another, conserving the properties of their latitude of origin. Stratosphere temperatures are high in the Arctic and we shall see that ozone concentrations are also high, while the reverse is true in both cases in the tropics. Thus the moving air-mass theory explains this correlation, although, as will be seen in §4.33, it is not so successful in other circumstances.

Temperatures measured from the contrast in the Huggins bands may vary very rapidly with time. For example, during the night of 1/2 March 1938, measurements on the Jungfraujoch indicated a change of 65° C. in 2¼ hr.

TABLE XIII. *Stratospheric temperatures from measurements on the Huggins bands*

| | | Ozone amount (cm. S.T.P.) | | | | |
		0·18	0·20	0·23	0·26	
Absiko	Temperature ° C.	−67	−44	−43	−33	−10
(Lapland)	Number of obs.	6	4	3	3	10
Arosa and	Temperature ° C.	−50	−42	−34	−24	−18
Jungfraujoch	Number of obs.	8	12	17	8	4
Total	Temperature ° C.	−57	−43	−35	−26	−12
	Number of obs.	14	16	20	11	14

4.242. *Emission temperature of the 9·6 μ band.* The 9·6 μ band in the solar spectrum is intense and is in a region where there are no other major absorptions; moreover, it is near to the peak emission of black bodies at normal terrestrial temperatures. This combination of facts makes it possible to measure the emission temperature of the ozone layer. The emission temperature (θ) is defined by the relation

$$e_\lambda = a_\lambda B_\lambda(\theta),$$

where e_λ is the energy emitted by the ozone layer in the wavelength range λ to $\lambda + d\lambda$, a_λ is the absorptivity of the whole layer, and B_λ is Planck's function for black-body emission. This temperature has a simple meaning only if Kirchhoff's law is obeyed and if the ozone is all at the same temperature, in which case emission temperature and gas-kinetic temperature are the same. It will be shown in Chapter VI that Kirchhoff's law should be valid in the stratosphere, and, in middle latitudes, the lower stratosphere, which contains most of the ozone, is approximately isothermal.

Adel (1947 b, 1949, 1950) has measured e_λ and a_λ separately from atmospheric absorption and emission spectra. For daytime measurements a solar spectrum is recorded by a spectrometer with a rock-salt prism and a thermocouple detector (fig. 42(a)). The sloping base-line, obtained with a glass shutter, is used, since glass will transmit the short-wave radiation scattered by the spectrometer optics, and the same spurious signal thus appears upon the base-line and the record.

The solar image is then moved away from the slit of the spectrometer, and an emission spectrum is recorded, viewing the sky at the same zenith angle in order to use the same atmospheric path as was used for the absorption spectrum (fig. 42(b)). The spectrometer conditions are the same for both records except that the amplifier gain is varied. It will be seen that in order to record the emission spectrum very wide slits have been employed, which have smoothed out the doublet structure of the band.

For night-time measurements, lunar spectra have to be used, but, since the intensity is so low, the spectrum obtained is a mixture of the lunar spectrum and the atmospheric emission spectrum. The two effects can, however, be separated and the coefficients determined as for the solar records.

The accuracy of this experiment is remarkably high. Adel has shown that in the region of the $9·6\mu$ band $B_\lambda(\theta)$ varies as θ^6. A 12 % error in energy measurement leads to a 2 % error in temperature (i.e. 5° K.), and this is probably the upper limit of the error of the experiment.

From March to July 1948 a series of 138 measurements was made on 51 days, and nine measurements were made on five nights at Almagordo, New Mexico. Measurements taken on the same day, within an hour of each other, showed slight differences, the average being 4° K. and the maximum 12° K., and these differences are considered by Adel to be significant. The surprising feature of the results is that all the 138 daytime measurements lie within the limits 227–243° K., while the nine night-time measurements lie within the limits 230–239° K., indicating a remarkable constancy of stratospheric temperature and very little variation

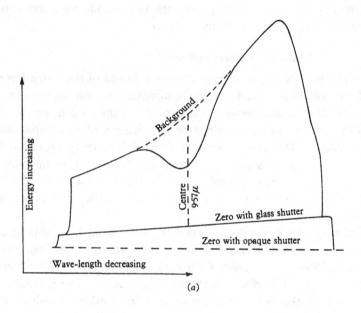

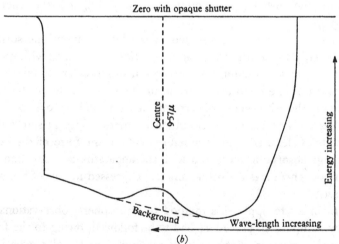

Fig. 42. Determination of the emission temperature of the $9 \cdot 6\mu$ band of ozone. (After Adel, 1949.) (a) Solar absorption spectrum; (b) thermocouple emission spectrum.

between day and night. The average temperature for the whole series is 235° K., which agrees with radio-sonde measurements for the level of 29 km. in this district.

4.25. *Mean height determination*

The methods of measuring the mean height of the ozone layer discussed in §4.221 suffer from the disadvantage that one measurement takes a considerable period of time; they are therefore of little use for measuring short-period changes of this important parameter. However, a method devised by Strong (1941) and Watanabe (1943) is capable of making an instantaneous measurement. This method depends upon the fact that the 9.6μ band is strongly pressure-dependent, whereas the Hartley and Huggins bands are not.

Strong and Watanabe observed the 9.6μ band by means of a residual ray apparatus in which successive reflexion from five apophyllite crystals isolated a narrow band of wave-lengths from 9.3 to 10.0μ. The efficiency of this apparatus as a filter is demonstrated by the fact that the scattering from other wave-lengths inside the instrument amounted to only $\frac{1}{4}\%$ of the energy at 9.6μ.

The apparatus was calibrated in the laboratory by measuring the energy which penetrated an absorption tube, sealed with rock-salt windows, containing ozone-oxygen mixtures at various total pressures. The quantity of ozone in the tube was found by observing the ultra-violet absorption at 3050 and 3110Å. with a Hilger-Müller double-quartz monochromator. Fig. 43 shows the results of these laboratory measurements in the form of the percentage absorption measured with the apparatus as a function of $\sqrt{(10x)}$, where x is the ozone amount expressed in cm. of the gas at S.T.P.

In order to apply this method to atmospheric observations a very complicated procedure had to be followed, owing to the fact that the response of the residual ray apparatus to solar radiation in the absence of the atmospheric ozone was not known. The response of the apparatus to sunlight could be observed over a large range of zenith angles and, by extrapolation, the response without any atmosphere could be inferred. However, the apparatus was not sufficiently stable to retain its calibration; moreover, water-vapour absorption also occurs in this region and can vary from day to day. The first difficulty could be overcome by standardizing the apparatus on a laboratory source. The second difficulty required measurements of the total water vapour in the optical path through the atmosphere. These were obtained by

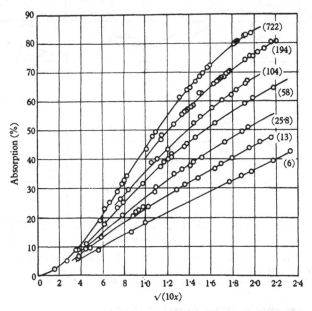

Fig. 43. Absorption by the $9 \cdot 6 \mu$ band as a function of amount and pressure. (After Strong, 1941.) The figures in parentheses give the total pressure in mm. Hg; x is the ozone amount in cm. at S.T.P.

observing the sharp ϕ band of water vapour at $1 \cdot 16 \mu$ and using laboratory observations from which it was possible to calculate the associated absorption at $9 \cdot 6 \mu$. The measurements in the near infra-red and ultra-violet spectrum were both made with the quartz monochromator, and the ozone amount was calculated by the methods described in §4.21.

Having measured the total ozone in the atmosphere and the absorption in the $9 \cdot 6 \mu$ region it is only necessary to consult the curves shown in fig. 43 to find the mean pressure of the ozone. Fig. 44 shows the results obtained on four days. The mean height during daytime was measured on 31 days and remarkably little day-to-day variation was observed. The average for all observations was 22·7 km. and the maximum variation 20·4–24·7 km., which could be attributed to errors of observation rather than to any changes in the atmosphere.

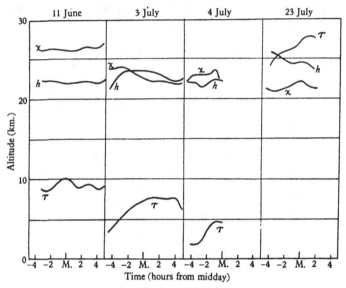

Fig. 44. The mean height of the ozone layer (*h*) together with the total ozone amount (χ) and the total water vapour (τ) in the atmosphere. (After Watanabe, 1943.) The units of the ordinate are: *h*, 1 km.; χ, 0·01 cm. at S.T.P.; τ, 0·1 g.cm.$^{-2}$.

4.3. Systematic ozone measurements

The total amount of ozone is the simplest parameter of importance which can be measured, and Dobson and his colleagues have collected data for many latitudes and at all seasons over the past 25 years. The main practical application of these measurements lies in the correlation which exists between short-period changes and weather systems, suggesting that a powerful synoptic tool has been discovered.

4.31. *Latitude and seasonal variations*

These are best discussed by means of the composite diagrams (fig. 45) drawn up by Craig (1950) and embodying all the information available before 1950. Some of the irregularities in these diagrams may be because all observers do not use the same technique or apparatus, and some because the observations may be insufficient to smooth out all the short-period variations. Nevertheless, the regularities are very striking. Near the equator there is very little seasonal variation in the total ozone amount, while the variation increases both to the north and to the south until there is an amplitude of 50% at 70°. The phase of the variation is the same in the two hemispheres, with a maximum

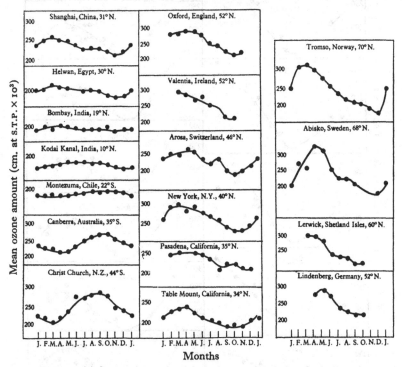

Fig. 45. (a) Seasonal variation of ozone amount. (After Craig, 1950.)

ozone amount occurring in the spring and a minimum in the autumn. There is also some evidence in favour of a subsidiary maximum occurring in late summer. The same results plotted with latitude as the abscissa (fig. 45(b)) show that at all times of the year there is a pronounced minimum of ozone amount in the tropics. There is also some evidence for a maximum of ozone amount near 60°, and therefore presumably for a minimum in the Arctic regions. In the past, observations near the poles have not been reliable owing to the difficulties of observation, particularly during the winter; it is clearly important that this gap should be filled.

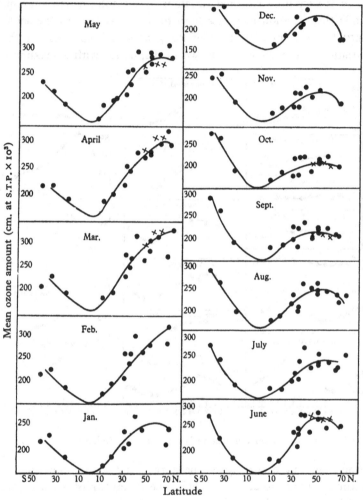

Fig. 45. (b) Latitude variation of ozone amount. (After Craig, 1950.)

4.32. *Correlation of ozone with other factors*

There have been a number of attempts to discover a correlation between ozone amount and other, possibly related, factors such as sunspot number, solar constant, magnetic phenomena, ionospheric storms, etc. No significant correlations have yet been demonstrated.

One investigation of great fundamental importance has been the search for a diurnal variation of ozone amount. Although such diurnal variations have been reported, the criticism of Dobson and Harrison (1926) has yet to be answered; they pointed out that, since the method of measurement assumes that no variation takes place on a settled day (see §4.21), it will be impossible to detect any regular effect which takes place on all days.

4.33. *Ozone and weather*

Weather relations are usually studied statistically or on a synoptic basis. The first method yields precise numerical results, for which any theory must account, but cross-correlations are often so many and so complicated that it may not help much in the formulation of a theory. The second method, although essentially descriptive, is probably the more effective from this point of view.

Statistical analyses have been performed by Dobson, Harrison and Lawrence (1927) and by Meetham (1936). Meetham managed to eliminate seasonal effects with some confidence, leaving only such correlations as result from the short-period changes. His work shows that there is a negative correlation of 0·5 between ozone amount and atmospheric pressures up to 18 km. There is also a positive correlation of 0·6 between ozone amount and the stratospheric temperature. The interplay of these two factors is such as to give a high positive correlation (0·75) between ozone amount and the potential temperature† in the stratosphere, and a high negative correlation between ozone amount and stratospheric density.

The relations which have been demonstrated between ozone amounts and weather phenomena are of theoretical importance, since the former are usually considered to be related to stratospheric conditions, and the latter to tropospheric conditions. This is a good example of the complicated and unexpected interactions encountered in most problems of atmospheric physics. Dobson

---

† Potential temperature is the temperature which the air would have if compressed adiabatically to a pressure of 1 atmosphere and is a measure of the entropy.

and others (1927) showed by synoptic methods that the ozone distribution near an anticyclone or a cyclone looks very similar to the ground-level pressure distribution, i.e. contours of constant ozone amount are almost circular, with a centre near to that of the pressure distribution. The pressure and ozone amount have a negative correlation, so that ozone amount is low over the centre of an anticyclone. The change in ozone amount from the centre to the edge of a typical anticyclone is about 25% or 0·07 cm. There may be a measurable displacement between the centres of the ozone and the ground-level pressure systems, but since the atmosphere is practically always sheared by winds, this is not surprising.

Similar relations exist between the ozone amount and the pressure distribution at frontal surfaces. Dobson and others (1946) remark that 'the passage of a warm front, when the cold air is replaced by warmer air, is accompanied by a fall in the ozone content, and since a warm front slopes forward at an angle of roughly 1/50, the fall takes place before the arrival of the front at ground level, just as the upper clouds show the approach of a warm front before the arrival of the front at the surface'. Similar remarks, but in reverse, apply to a cold front. Important relations between the ozone distributions and occlusions have also been found, but these are very complicated and will not be discussed here.

There have been many discussions on the question of whether a moving air mass preserves the characteristic ozone amount of its latitude of origin. This notion has already been mentioned in §4.241 in connexion with short-period changes of atmospheric temperature. It also accounts to some extent for the short-period ozone changes, since there is undoubtedly a tendency for air masses coming from higher latitudes to have a high ozone content, while those coming from lower latitudes frequently have a low ozone content. If the origin of air masses could be found by ozone measurements it would be of great practical importance in weather forecasting, but Dobson (1930) has pointed out two objections to this idea. He has shown that, at times, the ozone amount in England may be greater than the mean monthly amount at any other latitude, and that the day-to-day changes in ozone are nowhere near proportional to the latitude gradient. While both of these arguments are of weight, some of the evidence undoubtedly supports the theory that short-period changes are symptomatic of the conservative properties of air masses, although it is quite clear that there must also be other processes which are at least as important.

4.4. Theoretical treatment

4.41. *Introduction*

Oxygen can exist in the atmosphere in three forms, viz. molecular oxygen, atomic oxygen and ozone. The equilibrium between these three components is governed by the intensity of ultra-violet radiation from the sun, by the various reaction constants involved and, as will be seen later, by the degree of mixing in the stratosphere.

The first attempts to explain the vertical distribution of these three forms of oxygen were qualitative, but nevertheless Chapman (1930), Mecke (1931) and Wulf (1932) all reached the same general conclusions. Above 90 km. the particle density is very small, and reactions involving only one atmospheric molecule or atom are greatly favoured in comparison with those which involve two or more. High-energy quanta from the sun which are capable of dissociating oxygen molecules can penetrate to this level with only slight attenuation, and as a result the dissociative reaction between quanta and oxygen molecules predominates over the associative reaction between two oxygen atoms, and the oxygen is almost entirely in the atomic form. Ozone is not formed since a three-body collision involving an atom and a molecule of oxygen is required, and at such low densities this can be shown to be very improbable. At lower levels the probability of three-body collisions is much higher, and ozone will be formed at levels where this probability is high enough and a few high-energy photons can still penetrate. At still lower levels there are no high-energy photons which can dissociate the oxygen molecules, since they have all been absorbed by the oxygen and ozone in the higher layers. Thus the qualitative theory indicates that above 90 km. oxygen is atomic, while below this level it is molecular with a very slight admixture of ozone, which should have a maximum concentration in the stratosphere.

Quantitative treatments of the problem are fraught with difficulties and require very precise laboratory data. Calculations have nevertheless been made by Wulf and Deeming (1936), Schröer (1949), Dütsch (1946) and Craig (1950), and in the following section we shall follow mainly the work of the last-named author.

4.42. *The photochemistry of atmospheric oxygen*

The dissociation energy of molecular oxygen corresponds to a quantum of wave-length 2400Å., and if a molecule absorbs a quantum of higher energy the following reaction will take place:

$$O_2 + h\nu \rightarrow O + O, \quad \lambda < 2400 \text{Å.}, \quad \text{absorption coefficient } \alpha_{2_\lambda}.$$

Oxygen can absorb any energy below this wave-length in one of three major band systems, viz. the Herzberg, Schumann and Hopfield bands.

Ozone is less stable than molecular oxygen and may be dissociated by the absorption of quanta of wave-length less than 11,000Å., and therefore by any radiation absorbed in the Chappuis, Hartley or Huggins bands. The following reaction takes place:

$$O_3 + h\nu \rightarrow O_2 + O, \quad \lambda < 11,000 \text{Å.}, \quad \text{absorption coefficient } \alpha_{3_\lambda}.$$

It is possible to write five chemical reactions between the three forms of oxygen. Two of these represent coalescence of two particles into one, and a third body is therefore required to conserve energy and momentum. The third body (M) may be any air molecule or atom:

$$O_2 + O + M \rightarrow O_3 + M \quad \text{rate constant } k_{12},$$
$$O_3 + O \quad\;\; \rightarrow 2O_2 \quad\quad\;\; \text{rate constant } k_{13},$$
$$O + O + M \rightarrow O_2 + M,$$
$$O_3 + O_3 \quad\;\; \rightarrow 3O_2,$$
$$O_3 + O_2 \quad\;\; \rightarrow O + 2O_2.$$

It will be assumed that the last three reactions are unimportant. The rates of these reactions are not very well established, but all investigators agree that they are unimportant below 50 km. Above this level the recombination between two oxygen atoms should, however, be considered. If the results of Craig's computations in fig. 46 are compared with the rocket results given in §4.222, it will be realized that the observed concentrations lie significantly below the computed concentrations for altitudes above 55 km. Johnson and others (1952) have shown that this discrepancy no longer exists if the recombination between oxygen atoms is taken into account.

If n_1, n_2, n_3 and n_m are the numbers of oxygen atoms, oxygen molecules, ozone molecules and air molecules respectively per cm.^{3}, and if $q_\lambda d\lambda$ is the number of quanta between λ and $\lambda + d\lambda$ incident on a specified layer of the atmosphere, then, from the four reactions whose rates or absorption coefficients have been given, the following equations for the rates of change of oxygen atoms and ozone molecules can be written:

$$\frac{dn_1}{dt} = 2n_2 \int_{\lambda < 2400} \alpha_{2_\lambda} q_\lambda d\lambda + n_3 \int_{\lambda < 11,000} \alpha_{3_\lambda} q_\lambda d\lambda - k_{12} n_1 n_2 n_m - k_{13} n_1 n_3,$$
$$\quad\quad\quad (O_2 \to 2O) \quad\quad\quad (O_3 \to O_2 + O) \quad\quad (O_2 + O \to O_3)\ (O_3 + O \to 2O_2)$$

$$\frac{dn_3}{dt} = k_{12} n_1 n_2 n_m - \int_{\lambda < 11,000} \alpha_{3_\lambda} q_\lambda d\lambda - k_{13} n_1 n_3.$$
$$\quad\quad (O + O_2 \to O_3) \quad\quad (O_3 \to O_2 + O) \quad\quad (O + O_3 \to 2O_2)$$

If we now consider the equilibrium case, then both of these expressions are equal to zero, and by eliminating n_1 the following equation may be obtained for n_3:

$$n_3 = \frac{k_{12}}{k_{13}} n_2 n_m \frac{Q_2}{Q_3 + Q_2},$$

where $Q_2 = n_2 \int_{\lambda < 2400} \alpha_{2_\lambda} q_\lambda d\lambda =$ rate of destruction of oxygen molecules, and $Q_3 = n_3 \int_{\lambda < 11,000} \alpha_{3_\lambda} q_\lambda d\lambda =$ rate of destruction of ozone molecules.

This expression has to be solved by numerical methods, since it is far more complicated than may appear. Successive approximations have to be made, since n_3 occurs also on the right-hand side in the factor Q_3, and, moreover, q_λ involves n_3 at all levels above the one considered.

Up to the stratopause the values of n_2 and n_m are known from temperature and pressure measurements. The ratio k_{12}/k_{13} can be measured in the laboratory, where it is found that it depends critically upon temperature, being 150 times greater at -60 than at $+57°$ C. The atmospheric temperature is therefore an important variable, and since, as will be shown in Chapter VI, stratospheric temperature is largely controlled by the ozone concentration, there is here a complicated interdependence of phenomena which is typical of atmospheric problems. The greatest difficulty in the calculation of the equilibrium case is the evaluation of Q_2, since the oxygen absorption coefficients in the ultra-violet spectrum are none too well determined and they depend markedly upon the total pressure in a way which is not well understood. Furthermore, the important region near 2000Å. is also overlapped by the Hartley band, and so Q_2 depends, to some extent, upon the total ozone above the level concerned. On top of all these difficulties, the solar radiation incident on the atmosphere from 3000 to 1000Å. is not known with certainty, and this quantity is involved in both Q_2 and Q_3.

4.43. *Results of calculations*

Fig. 46 shows the vertical distribution of ozone as computed by Craig, assuming equilibrium with a zenith sun and that the oxygen absorption coefficients near 2000Å. do not depend upon pressure; this is compared with inversion-effect measurements at Arosa. The total amount calculated is 0·298 cm. of ozone at S.T.P., and the maximum concentration lies at 30 km. Both of these figures are in good agreement with observation. Above 35 km. the theoretical and observed curves agree very well, but when the difficulties of interpreting the inversion effect are remembered this is not of great significance. Below 20 km. the two curves in fig. 46 diverge rapidly, and it appears that no reasonable alterations of the constants involved will explain the large ozone concentrations observed in the troposphere.

Craig also investigated the ozone distribution making reasonable allowance for the known effect of pressure on the oxygen absorption. Compared with the calculations made without including this effect, the ozone values above 30 km. were increased, while below this level they were decreased; the calculated total ozone increased by a factor 2. The importance of the precise temperature structure of the stratosphere, through its influence on the rate coefficients, was assessed by making calculations for a range of possible temperatures. Despite the critical dependence of the rate coefficients upon temperature, Craig found changes of only 10% in the total amount of ozone for large changes in the temperature structure.

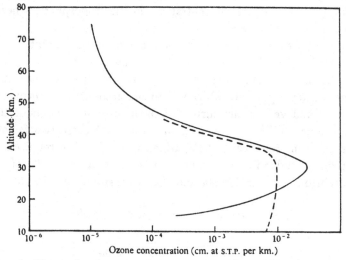

Fig. 46. Observed and computed ozone concentrations. (After Craig, 1950.) - - - - inversion effect at Arosa; ——— calculated assuming no pressure dependence of oxygen absorption coefficients.

It has been suggested that the observed latitude and seasonal changes in ozone are bound up with the variation of solar zenith angle. Calculations under otherwise comparable conditions gave the following values for the total ozone amount in equilibrium with a solar beam at various zenith angles: zenith angles, 0, 22, 45, 67°; ozone amounts, 0·298, 0·262, 0·196, 0·085 cm. at s.t.p. Practically the whole of the changes involved take place below 35 km.; above this level the zenith angle appears to be un-important. Since the ozone amounts are least in the tropics, where the zenith angle is least, this effect indicates the reverse of the observed latitude variation. Since, at all latitudes, ozone has its local maximum in spring, for intermediary zenith angles, it also cannot explain the observed seasonal variation of ozone amount.

4.44. Comparison with observation

The equilibrium calculations described in the last section give a good account of the mean ozone amounts and of the concentration maximum which is observed in the stratosphere. They fail entirely to account for the following observations:

(a) the large ozone concentrations in the troposphere, where, according to the theory, very little is formed,

(b) the latitude and seasonal variations of ozone amount,

(c) the association of ozone with weather systems.

The most obvious weakness in the theory as it stands is the equilibrium hypothesis, which fails to take account of the turbulent mixing and vertical air currents known to exist throughout the stratosphere (Chapter V). The importance of these two factors can be seen from Table XIV, which gives the time required at various altitudes for the ozone amount to decay to $2 \cdot 6\%$ of its equilibrium value, if the solar radiation were suddenly to be cut off.

TABLE XIV. *Ozone decay times at various altitudes*

Height (km.)	Time of decay to $2 \cdot 6\%$ of the equilibrium concentration (days)
38	0·1
35	1
32·5	1–10
30	10
27·5	100
25	100–1000
22·5	1000
20	10,000

Only orders of magnitude are given in Table XIV, and they are roughly representative of all the various conditions which have been used by Craig to compute the ozone amount. It is impossible to say exactly where the probable boundary between equilibrium and non-equilibrium must lie, but it is reasonable to surmise that changes which take longer than one day will never come into equilibrium, while those which take place in a day or less may have this opportunity. If this is correct, then there is a very clear division near the calculated maximum ozone concentration, above which the ozone may be in photochemical equilibrium, but below which it behaves more like a permanent constituent of the atmosphere (that is, neglecting catalytic decomposition by dust and by the earth's surface, which we have not discussed). Thus the region above the ozone maximum may be considered to be an inexhaustible source or sink of ozone, while the region below may be considered to be a reservoir for the gas. A downflow of air would transfer ozone from the upper region, where it would immediately be replaced, to the lower region, where it would be stored, giving a net gain of ozone entirely in the lower region. Similarly, an upflow of air will cause a decrease in the net ozone amount, mainly in the lower region.

Since, in the region below the ozone maximum, the ozone concentration increases with height, the effect of turbulent mixing will be to transfer the gas downwards,† also leading to a net increase in the ozone amount.

These factors account qualitatively for the difficulty that the ozone concentration is higher than the equilibrium value in the troposphere. They appear to indicate that changes in total ozone amount should take place mainly below the level of maximum concentration, and the results shown in fig. 43 suggest that this is so. They also hold out hopes of explaining the connexion between ozone and weather systems through the degree of mixing which takes place in the higher layers. Quantitative estimates are difficult to obtain, although Nicolet (1945) has been able to show that the effects caused by mixing are of the right magnitude to explain observed short-period changes of ozone amount. Reed and Julius (1951) have calculated the magnitudes of the vertical velocities and the turbulent exchange coefficients required to explain the observed seasonal changes of ozone between 10 and 20 km. They find downward velocities of about 1 m./sec. or exchange coefficients of 10 g.cm.$^{-1}$ sec.$^{-1}$. According to ideas from

† The elementary theory of turbulence demands that transfer takes place down a gradient of mixing ratio.

other sources these values are rather large (see Chapter V), but they cannot be excluded from consideration. Reed (1950) has given qualitative explanations of nearly all the known ozone and weather relationships in terms of the field of flow in the lower stratosphere and its effect on ozone concentration.

These developments suggest that the whole ozone problem must be treated in terms of non-equilibrium processes. This means that final solutions are a long way off, since they demand an exact knowledge of the field of flow and the state of turbulence in the stratosphere.

CHAPTER V

WINDS AND TURBULENCE

The subject of the hydrodynamics of the stratosphere is growing to be one of importance to synoptic meteorology. The idea is now gaining ground that, from the dynamical standpoint, the stratosphere and the troposphere should be treated as a single entity. We do not presume to give an adequate treatment of the subject from this point of view, but will simply discuss such features as appear to be of importance in considerations of heat transfer.

It is, of course, not exactly clear which factors are of importance, but it can at least be stated that there is an intimate relation between manifestations of kinetic energy and the transfer of heat. Winds can only be driven if heat is degraded from higher to lower temperatures, and turbulent mixing will transfer heat in any atmosphere which has not an adiabatic lapse rate. Although winds and turbulence are both part of the same process, they are usually distinguished, and for this reason they will be separately treated here, although, as will be seen, the distinction is often rather arbitrary.

5.1. Stratospheric winds

5.11. *Near the tropopause*

In 1934 it was still possible to publish the following statement (Crossley, 1934): 'The stratosphere appears to be an inactive region (as regards heat) in which nothing much happens.' It is not suggested that this statement represented the generally accepted view even at that time, since Dobson (1920) had measured considerable wind strengths in the lower stratosphere, and this suggests the possibility of turbulent transfer. Dobson's results were obtained with sounding balloons tracked by theodolites, and, as might be expected, they varied considerably from day to day. One result stands out clearly, however, particularly with the stronger winds: wind speed generally increases with height in the troposphere and there is a sharp maximum at the tropopause, above which it decreases with height, certainly as far as 14 km. There is no comparable change in wind direction as the tropopause is crossed. It should be remarked that, for seasonal averages, such as those shown in fig. 49, the maximum wind speed at the tropopause is not so very marked, although individual ascents show the effect quite strikingly.

More recently it has been realized that the high winds at the tropopause are restricted to a comparatively narrow belt of latitudes. Some hint of this will be found in the well-known textbook *Physikalische Hydrodynamik* (V. and J. Bjerknes, Solberg and Bergeron, 1933), where it is shown that the computed gradient wind† in the northern hemisphere during summer can be very strong at the tropopause, but only in middle latitudes.

In the northern hemisphere this concentrated high wind near the tropopause is westerly, and attention has only recently been focused upon it by the Chicago school of meteorology (University of Chicago, 1947), who have coined the descriptive term 'jet stream'.

Fig. 47 shows a meridional cross-section of most of the northern hemisphere running through America, with observed mean temperatures and computed wind speeds. The tropopause is drawn with a break at 35–40° N., and at this break there is a very strong computed geostrophic wind (westerly, stronger than 180 km./hr.), which falls off above and below and to the north and to the south. The jet stream is therefore a tongue of high-speed westerlies some kilometres thick and with a core about 10° of latitude wide. If specific examples rather than average figures are analysed, then the jet is even more concentrated than is shown in fig. 47.

Other characteristics of the jet stream are its patchiness and its tendency to meander around the pole in huge, but quite regular, waves. The patchiness is illustrated in fig. 48, where, once again, averaged data are used. Despite the averaging, the maximum velocities (about 200 km./hr.) are quite impressive.

The wave structure of the jet stream is only clear if averaged

† The gradient wind is the wind whose centrifugal and Coriolis forces (caused by the earth's rotation) just balance the pressure gradient; there is usually good agreement between the gradient wind and the observed wind in the troposphere, even when the centrifugal term is omitted (giving the geostrophic wind).

data are not used. The only published example of a hemispherical analysis of this nature has been given by Rossby (1949) for the 500 mb. level. This analysis shows a marked westerly jet circulating around a mean latitude of 60° N., but with deviations from the mean in the form of three enormous waves which give the synoptic chart the appearance of a crude clover leaf.

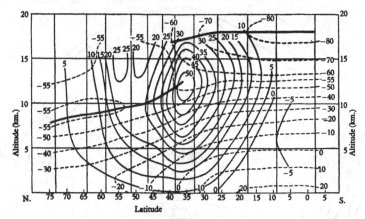

Fig. 47. Observed mean temperatures and mean geostrophic winds computed from the observed temperature and pressure data in a vertical north-south section through Central and North America. (After Rossby, 1949.) Based upon daily radio-sonde data (vertical lines show the positions of the observing stations) for January and February 1941–5. The heavy line represents the tropopause; full lines give wind speeds in m.sec.$^{-1}$; dashed lines give isotherms in degrees C.

5.12. In the lower stratosphere

Our information about winds above the tropopause is almost entirely restricted to temperate latitudes in the northern hemisphere. Here the predominant winds in the upper troposphere are westerly (Brooks, Durst, Caruthers, Dewar and Sawyer, 1950), and the change in passing through the tropopause is in magnitude and not in direction. As far as radio-sonde measurements go this picture is not greatly changed except that the westerly wind at 20 km. is much weaker in summer than in winter.

Fig. 49 shows mean curves for England and Scotland, and the calculated curve is obtained from a variant on the gradient wind equation which relates the change in wind velocity from top to bottom of a layer to the temperature gradient within the layer. Theory and measurement both seem to indicate that above 20 km.

the wind changes direction from easterly in summer to westerly in winter—the monsoon effect,† so called from its obvious parallel.

The existence of a monsoon wind at high levels was suggested by Whipple (1935) to account for the seasonal variation of the zones of audibility for explosions in North-West Europe. In the

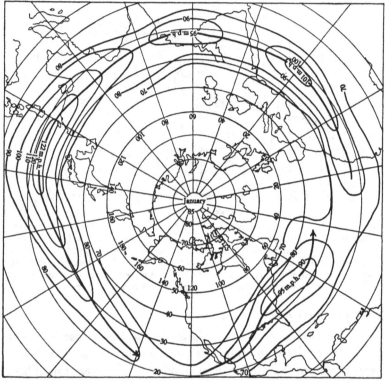

Fig. 48. Geostrophic wind speeds in m.p.h. at 10 km. in the northern hemisphere. Average for several Januaries. The thick line gives the locus of the maximum wind. (After Namias and Clapp, 1949.)

winter the abnormal audibility is better to the east than to the west, and in summer the position is reversed. Since it is known that sound-ray trajectories reach to 40 km. it was suggested that this indicated winter westerlies and summer easterlies near 30 km. Carefully planned sound-propagation experiments by Crary

† By definition, a monsoon wind blows in opposite directions in summer and winter.

(1950) (see §2.3 and fig. 16) have allowed the separation of wind from temperature effects, showing that this suggestion is entirely correct.

In order to confirm his deduction, Whipple suggested that measurements should be made of the motion of the burst of a gun shell fired into the stratosphere. This experiment was performed by Murgatroyd and Clews (1949), using a Bruce gun firing smoke shells to 30 km. The monsoon wind was very clearly observed.

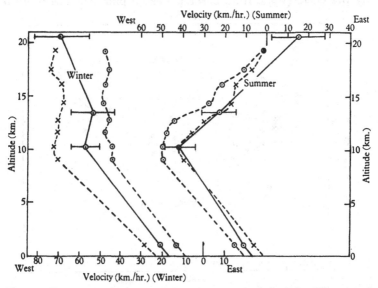

Fig. 49. East-west components of the mean vector wind. (After Murgatroyd and Clews, 1949). - - - × - - - Lerwick (Shetland Islands); - - - ⊙ - - - Larkhill (Wiltshire); —⊙— calculated mean from the horizontal temperature difference between the two stations (horizontal lines give possible errors).

From October to April the winds were westerly, with a maximum speed of 180 km./hr.; from May to September, however, light easterlies were observed, with a maximum speed of 50 km./hr. in July. As far as could be judged these winds agreed well with the predicted gradient winds, although temperatures above 20 km. had to be assumed.

Recently it has been possible to make reliable routine ascents to 30 km. using large rubber balloons, and in the next few years we may look forward to reliable data about the general circulation of the lower stratosphere. Scrase (1951) has completed one year's observations over Great Britain up to 30 km., and has been able to determine how the height of the transition from easterlies to westerlies varies with season. Fig. 50 shows Scrase's results. In summer the transition level is at 20 km., which confirms the other sources of information, but in winter it lies above 30 km., and the indication is that it may be very far above this level. Less detailed results by Crary (1950) from sound-propagation experiments and by Brasefield (1950) from observations on pilot balloons confirm

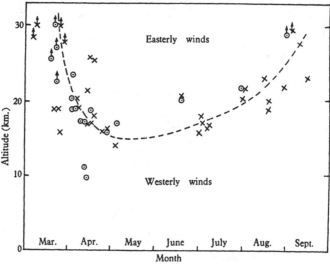

Fig. 50. The transition height from easterly to westerly winds over Great Britain throughout the year. (After Scrase, 1951.) ⊙ Lerwick (Shetland Islands); × Downham Market (Norfolk).

these results, and, in addition, Crary's work shows that, throughout the lower stratosphere, diurnal wind components are small (see §2.3).

5.13. *In the upper stratosphere*

The following phenomena and methods have been used to obtain the winds at levels above 30 km.:

    (*a*) sound propagation,
    (*b*) persistent meteor trains,
    (*c*) noctilucent clouds,

(d) smoke and vapour trails from V2 rockets,

(e) movement of ionized clouds,

(f) horizontal temperature gradients,

(g) theoretical work on atmospheric oscillations and the dynamo theory of quiet-day magnetic disturbances.

Before describing some of these observations it is best to summarize the findings, since the detailed results are somewhat contradictory.

(i) Wind speeds are invariably great in the upper stratosphere (although it should be realized that the energies involved may be quite small).

(ii) Experiments do not indicate a prevalence of east-west winds at all levels, as in the troposphere and lower stratosphere. There are, however, east-west components of considerable strength, and these vary seasonally having a monsoon character. The summer winds are easterly below 60 km., but precisely the reverse at higher levels, with a transition somewhere near the stratopause.

(iii) Rapid and irregular changes of velocity are common, and semi-diurnal and diurnal components, which are small compared with the mean wind below 60 km., appear to be measurable at higher levels.

We will now describe the various observations which have been made, and it will be seen how far they confirm the above statements.

Crary's sound-propagation experiments at three latitudes have already been mentioned, and his results up to 60 km. are shown in fig. 16 (a) and (b). These diagrams suggest that only east-west winds were observed, but this is not so; the magnitude of the wind is plotted and the words 'easterly' and 'westerly' indicate only that the azimuth was less than or greater than 180°. The azimuths measured did not vary very much with height, and the average values were 88° for Bermuda, 100° for the Canal zone, 151° for Alaska (summer) and 202° for Alaska (winter). The last azimuth was very difficult to determine, and the results in fig. 16(a) were reduced on the assumption that the wind was due west. Both the Alaska experiments therefore indicate strong southerly components, although for lower latitudes north-south components were much smaller.

In general, Crary's results indicate that the monsoon wind which first appears above 20 km. intensifies up to about 50 km., where the maximum velocities are 220 km./hr. (Bermuda), 190 km./hr. (Canal zone), 70 km./hr. (Alaska, winter) and 220 km./hr. (Alaska, summer). All results indicate a decrease of velocity again after the maximum has been reached, suggesting that somewhere near 70 km. there may be a reversal of wind direction at all seasons.

Another important result obtained by Crary, which has already been mentioned in §2.3, was that up to 60 km. he could not obtain results which might indicate a substantial diurnal variation of the wind.

The use of V2 rockets for investigating stratospheric winds is in its earliest stages, and as yet only one useful observation has been published, showing a speed of 300 km./hr. at 30–35 km. in December 1948. Rockets provide the best potential means for systematic investigations, and we may look forward to the time when smoke-puffs are regularly released from high-altitude rockets and tracked from ground stations.

Information about meteors has been carefully collected by astronomers for more than 200 years. Those meteors which leave a persistent train can be used as tracers for the atmosphere between 30 and 120 km. Observations may be made visually or photographically, or, as in recent years, by reflexion of radio waves from those trains which are ionized. The material available for analysis is now immense in bulk, and Olivier (1947) has been able to publish figures for the drift of 1500 trains. He does not attempt to analyse them into height ranges, presumably because the height determinations are very rough, and the information therefore takes the form of averages over all seasons for levels somewhere in the stratosphere at some time during the night. These averages show a most complicated state of affairs, with apparently preferred directions at many different points of the compass, which vary, for example, with the time of day. Average velocities appear to be about 200 km./hr., and one isolated result as high as 7000 km./hr. has been reported. Some of the qualitative observations are of interest. For example, the trains often assume highly irregular shapes after a minute or two, showing considerable stratification of the wind, and there is some unmistakable evidence for the existence of appreciable vertical velocities.

Radio measurements of ionized meteor trains, which can be made at all hours of the day and night, have not yet disentangled this evidence. Fig. 51(a) shows the results of some recent measurements giving comparatively low speeds mainly in north-south directions (Manning, Villard and Petersen, 1950).

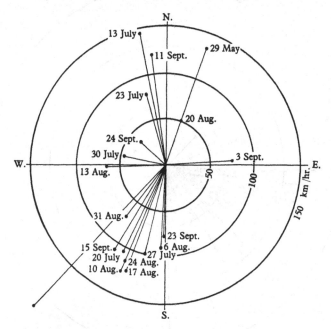

Fig. 51 (a). Vector diagram of winds in the upper stratosphere. 80–100 km. from ionized meteor trains. Average values from 2 to 4 a.m. local time during 1949. (After Manning and others, 1950.)

The noctilucent clouds, which are seen near to 80 km., are of great importance as tracers of atmospheric motions. These clouds are not observed so commonly nowadays as they were from 1885 to 1895, when they were carefully observed by Jesse (Archenold, 1928). More recent observations have been made by Störmer (1933), using his network of auroral stations. Vestine (1934) has examined all the observations before his paper, and his general conclusions are worth repeating. 'Velocities reached by the clouds are often truly remarkable. Jesse obtained a velocity as high as

625 km./hour in 1888. Before midnight most movements are from an azimuth 48° E., after midnight from an azimuth 63° E. with a possible secondary maximum from 73° W. The mean velocity for west azimuths was 142 km./hour and 241 km./hour from east azimuths. Movements from west azimuths were uncommon and

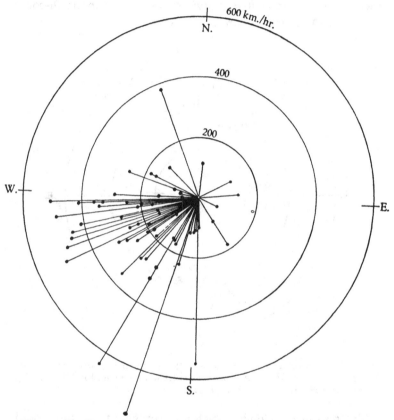

Fig. 51 (*b*). Vector diagram of winds in the upper stratosphere. 80 km. from noctilucent clouds. (After Vestine, 1934.)

few measurements show any southern component. Störmer found velocities of 150 to 200 km./hour from the N.N.E. for the clouds of July 10 to 11, 1932. Thus the clouds appear to move from the N.E. before midnight, from the N.N.E. after midnight and apparently with a component away from the position of the sun or the twi-light circle.' Fig. 51(*b*) shows a summary of measurements of noctilucent clouds after Vestine. There is little agreement either in magnitude or in direction between the measurements shown in fig. 51 (*a*) and (*b*), but this may be due to the fact that they are not strictly comparable.

Given the relation between temperature and height and the mean molecular mass for air, it is possible to calculate the gradient wind. Such methods give reliable results in the lower atmosphere, and Kellogg and Schilling (1951) have attempted to apply them to the upper stratosphere. As yet the temperature-height relations are but poorly established, but in the future measurements will probably become far more reliable. Even so, it is doubtful if they will ever provide a detailed explanation of upper stratospheric winds, since there are many indications that a wind system with very strong diurnal and semi-diurnal components is one of the main features of the upper stratosphere, and it will be a very long time before measurements throughout the day will be made anywhere except at ground level.

It has already been mentioned that observations of meteor trains and noctilucent clouds indicate some correlation with the time of day, suggesting wind components with diurnal or submultiple periods. There are several lines of evidence which suggest strong semi-diurnal components for the wind above the stratopause, and for the sake of continuity we will digress slightly in order to mention them.

Measurements of the fading of the intensity of radio waves reflected from the E region can be interpreted in terms of the movement of ionized clouds between 100 and 120 km. It is not yet completely certain that these motions are those of the air itself, although most objections to this idea seem to have been answered. Phillips (1952) has recently completed a series of observations in England, which have been confirmed by similar observations in Australia and the United States when referred to the local time. The average magnitude of measured winds is 290 km./hr., with peak values as great as 700 km./hr. Of this average magnitude about 40 % consists of a semi-diurnal component, while the remaining 60%, which is the same as the daily average wind, shows a seasonal variation of direction, being easterly in winter and westerly in summer. This result for the seasonal variation of the daily average wind supports very well the indication from the sound-propagation experiments that the summer easterlies only exists below about 70 km. and that above this level there is a complete reversal of the east-west components. The diurnal component, on the other hand, is not indicated by the sound-propagation results, which suggests that the ratio of the diurnal to the mean wind must increase between 60 and 110 km. by a fairly large factor.

Semi-diurnal wind systems in the E region of about the measured magnitude have been predicted from the theory of atmospheric oscillations (Pekeris, 1937), and are also invoked to satisfy Balfour Stewart's dynamo theory (1882) of the quiet-day variations in the earth's magnetic field. The phase of the measured winds differs, however, by 180° from that required by the dynamo theory. More exact temperature data are required in the E region before the phase of the semi-diurnal atmospheric oscillation can be determined precisely, but it seems likely that the oscillation has a measurably large wind component. Pekeris (1937) has worked out a numerical example which showed that in a particular case the velocities associated with the semi-diurnal oscillation were mainly horizontal, that they were fairly constant up to 30 km. where there was a node, and that above this level the velocities rapidly increased to 200 times greater at 100 km. The ground-level semi-diurnal wind is about 1·5 km./hr. in temperate latitudes, which is insignificant compared with the mean geostrophic wind. Pekeris's example suggests that up to 30 km. and over part of the upper stratosphere, the observed wind will have only small semi-diurnal components, but that above the stratopause they may form a large proportion of the observed mean. This is, of course, substantially what is observed, but more accurate observations and more precise computations are required before it can be confidently asserted that this feature of the problem is explained.

5.2. Turbulence in the stratosphere

Let it be stated at the outset of any discussion of turbulence in the stratosphere that our knowledge is very rudimentary and, at present, only qualitative discussion is possible. The problem is introduced only since it is of such great importance to the study of the thermal balance, and all that can be done is simply to indicate the various possible sources of information.

Our knowledge of atmospheric turbulence, even in the more accessible lower layers, is very incomplete and there is very little guiding theory. Nevertheless, it is certain that the following observations would be useful and they are all to some extent possible:

(a) The turbulent energy and its spectral distribution (i.e. the partition of the energy amongst disturbances of various sizes).

(b) The degree of anisotropy of the turbulent motions.

(c) Diffusion effects.

From (a) and (b) we might hope to learn about the energy balance in the atmosphere, and (c) is important because turbulent diffusion has an influence both upon heat transfer and upon the distribution of gases in the stratosphere.

5.21. *From stratospheric winds*

Irregular wind components can be looked upon as manifestations of large-scale turbulence.† Such large-scale turbulence is therefore indicated by most of the stratospheric wind observations, and we know that the large eddies will decay into smaller eddies until they reach such a size that viscous dissipation becomes large and the spectrum is cut off. To gain an idea of the extent of this spectrum, we may use a widely accepted expression (see, for example, Heisenberg, 1948) which relates the largest size of eddy (L_0) to the smallest size (L_s) in isotropic turbulence,

$$\frac{L_s}{L_0} \simeq 6 \left(\frac{v_0 L_0}{\nu} \right)^{-\frac{3}{4}},$$

where ν is the kinematic viscosity and v_0 is the root-mean-square velocity fluctuation. Suppose, for example, that the following values apply to the 50 km. level:

$$v_0 = 3 \times 10^3 \text{ cm.sec.}^{-1} \text{ (100 km./hr.)},$$
$$L_0 = 10^7 \text{ cm. (100 km.)},$$
$$\nu = 10^2 \text{ cm.}^2 \text{ sec.}^{-1},$$

we obtain $L_s = 30 \text{ cm.}$

† It should be realized that normal observational techniques lead to an apparent spectrum which is restricted at both ends; the largest eddy size recorded will depend upon the time of averaging used to obtain the mean wind, while the smallest eddy size will depend upon the frequency of the observations.

Thus, even though they are not directly observed, we may infer the existence of dissipating eddies with the scale of the order of 1 m.

An analysis has been made of the large-scale eddies as indicated by radio-sonde wind measurements (Brooks and others, 1950). The purpose was essentially practical in that it was desired to record the mean wind all over the globe and the probable departures from it. It was found that the vector deviation from the mean wind was of random direction and in magnitude followed the Gaussian distribution, which allows it to be specified in terms of one parameter only—the variance. The variance is not only a convenient parameter but it also gives a direct measure of the turbulent energy per unit mass. It appears that the variance increases with height through the troposphere, reaching a maximum about 2 km. below the tropopause; thereafter it decreases with height, but only to half the maximum value at 15 km. It would be most instructive if this analysis could be extended to higher levels.

5.22. *Meteorites and noctilucent clouds*

It has been mentioned that, after a minute or two, meteor trains often assume very irregular shapes, indicating large vertical gradients of the horizontal wind strength. It is possible in small-scale laboratory experiments to obtain stable laminar flow with considerable velocity gradients. In the stratosphere, however, the scale lengths involved are so great that hydrodynamical theory suggests that any measurable wind gradients will cause instability and therefore that the atmosphere will be turbulent.

It has also been mentioned that the observations upon meteor trains sometimes show the presence of vertical velocities, which likewise suggests a state of turbulent motion.

Noctilucent clouds give evidence about the atmosphere very near to 80 km. only. Observations indicate a complicated state of affairs, best described in the words of Vestine (1934): 'They [noctilucent clouds] are subject to rapid and continuous changes of shape and form, individual details in structure escaping recognition after an interval as short as five minutes. They may appear to twist about, and sometimes take on appearances like that of wound-up lengths of loose yarn. Occasionally they are made up of long parallel silvery streaks aligned along their direction of motion, with smaller parallel streaks at right angles filling the space in between and arranged in uniform wave-like sequences. These smaller clouds Jesse found at right angles to their direction of motion....Jesse obtained measurements varying little from a mean value of 9·4 km. for the distance between successive wave-like crests in the clouds, Störmer finding a value of 9 km. in 1932.'

The 'rapid and continuous changes of shape and form' indicate active turbulence at these levels on these occasions. The wave-like structure is a familiar phenomenon in the lower atmosphere and is usually taken to indicate cellular convection modified by wind shear.

5.23. Aircraft and balloon measurements

Aircraft flying in the stratosphere sometimes encounter localized patches of 'bumpiness' (Bannon, 1951). Much attention has been directed towards the explanation and prediction of this phenomenon, but no notable success has yet been achieved. The problem is a difficult one, because large atmospheric eddies take a very long time to decay, and the release of potential energy and the observed turbulent energy may be well separated in space and in time.

Observations have been made on the rate of ascent and the acceleration of pilot balloons (Koschmeider, 1936; Junge, 1938), and both methods indicate that the vertical component of the turbulent velocity may be greater in the lower stratosphere than in the troposphere. It is very difficult to interpret this kind of measurement, but it is possible that very interesting information about the anisotropy of the turbulent motions of the stratosphere could be obtained.

5.24. Diffusion measurements

If there were no mixing in the atmosphere each of the gaseous components would take up diffusive equilibrium with a scale height which depends on the molecular weight of the gas concerned. This would mean that the volume proportion of each gas would change very rapidly with height, the lighter gases becoming predominant as the height increases. With thorough mixing, on the other hand, the scale height for each gas would be the same, corresponding to a mean molecular weight for all constituents. For an intermediate distribution it is possible in principle to calculate the degree of mixing. This has been attempted by Lettau (1947) and Brewer (1949), who assumed that turbulent diffusion can be specified by means of an equation precisely similar to that used for molecular diffusion except that the coefficients differ. This is an acceptable assumption if the eddies responsible for the diffusion are very small compared with the scale of the phenomenon to be investigated. Calculations by Penndorf (1947), based on the data upon the vertical distribution of oxygen and helium available before the recent rocket experiments, give values for the factor

$$Q = \frac{\mu}{A + \mu}$$

as a function of height, where μ is the molecular diffusion coefficient and A is the corresponding turbulent coefficient. Penndorf's results are shown in Table XV. Brewer's calculation leads to smaller values of Q (i.e. larger turbulent diffusion) than those shown in Table XV.

TABLE XV. *Turbulent diffusion coefficients in the stratosphere*

Height (km.)	Q
20	0·05
30	0·10
35–40	0·15
40–50	0·25

In §3.4 it was pointed out that a possible cause of the observed helium surpluses in the lower stratosphere is a slow exchange of air between the levels of observation and levels above 70 km. If this is correct, the eddy sizes involved may be so great as to invalidate any argument based upon a simple turbulent diffusion equation. As a result the values given in Table XV may not be significant even as rough magnitudes.

5.25. *Theoretical considerations*

Between 60 and 80 km. there is probably a large lapse rate of temperature. The precise value of this lapse rate is uncertain, but some measurements suggest that it may be as large as 12–15° K./km. The adiabatic lapse rate at these levels is 9·7° K./km., and therefore if these figures are correct they indicate that the atmosphere must be in a state of active convection with a considerable upward heat flux. Kellogg and Schilling (1951) have tried to show from considerations of the radiation balance that such a heat flux is required to maintain a local equilibrium. This is a fairly powerful argument in principle, and it is probable that examination of the radiative heat balance will prove a useful weapon in the investigation of convective instability. In this case, however, the data used to compute the radiation balance are so speculative that one cannot place sufficient confidence in the result to do more than note that at least it fits in with other evidence.

CHAPTER VI

RADIATION

6.1. Introduction

6.11. *The radiation fluxes*

The ultimate source of thermal energy for the earth's atmosphere is the incident solar radiation, and we must know its precise form outside the limits of the atmosphere. Most observations take place at ground level, and, since the atmosphere absorbs and scatters a proportion of the incident radiation, our knowledge upon this subject is not direct. The total radiation received per unit area perpendicular to the sun's rays outside the atmosphere can be inferred from ground-level measurements made over a wide range of solar zenith angles by means of an extrapolation to the artificial case where $\sec \psi_s = 0$. The accepted mean value for the solar constant is $1 \cdot 95$ cal.cm.$^{-2}$ min.$^{-1}$ (Houghton, 1951), although the difficulties of measurement are such that this value may be in error by 2%. Owing to the eccentricity of the earth's orbit, the solar constant at the earth's surface will vary by $\pm 3 \cdot 5 \%$ from this mean value. The incident solar radiation per cm.^{2} of the earth's surface (which is the important factor meteorologically) is very variable, depending strongly upon season and latitude in a manner which may quite readily be computed when the solar constant is known.

It has been claimed by the observers at the Smithsonian Institute that the solar constant varies from day to day by 1 or 2 %. This is, however, scarcely outside the limits of error and cannot be said to be well established, although small changes must be expected in view of the observed changes which take place in solar topography.

It is not possible to establish with certainty whether or not the solar constant varies on a geological time scale, but such variations might reasonably be expected, and they are a prominent feature of the 'accretion theory' of stellar behaviour as developed by Hoyle (1949) and his colleagues.

Turning now to the spectral distribution of this energy, the problem becomes more complicated. In the visible spectrum the colour temperature of the sun is approximately 6000° K., although there is not a very exact correspondence with black-body radiation owing to the many Fraunhofer lines which cross the solar spectrum, caused by absorption in the solar atmosphere. Information from the ultra-violet region of the spectrum has become available in recent years by means of spectroscopic observation from rockets

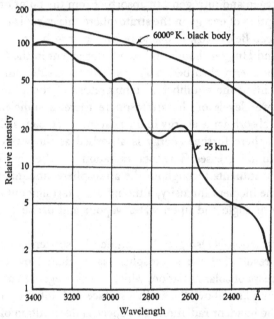

Fig. 52. The spectral distribution of solar energy in the ultra-violet as recorded at 55 km. (After Durand, 1949.)

above the ozone layer (Durand, 1949). These observations indicate a crowding together of the Fraunhofer lines in such a way that the smoothed-out contour of the solar energy curve falls below that appropriate to a 6000° K. black body.

The curve shown in fig. 52 corresponds more closely to a colour temperature of 5000° K., and even lower temperatures have been given for wave-lengths less than 2000 Å. While the normal colour temperatures in the ultra-violet may therefore be less than in the visible, there is indirect evidence that during disturbed conditions on the sun, such as large solar flares, vast excesses of short-wave radiation may be emitted.

In the infra-red region of the spectrum, measurements are few owing to the difficulties of the technique and the unsteady atmospheric absorption. Adel (1939) has given a value of 7000° K. for the colour temperature out to 13 μ.

The nearly parallel stream of solar radiation passes through the ionosphere with very little change except in the far ultra-violet where oxygen and nitrogen can absorb. From the point of view of the absorption of energy in the stratosphere this effect is not of any importance. Below 50 km. ozone begins to absorb strongly in the Hartley and Huggins bands and to a lesser extent in the Chappuis bands. The energy absorbed in this way is of the greatest importance in determining the equilibrium temperature of the stratosphere. In the lower levels of the stratosphere there is sufficient water vapour to absorb some energy in the near infra-red bands. Through the troposphere further energy is absorbed as the water-vapour concentration increases. The total radiation absorbed by all atmospheric constituents throughout the atmosphere amounts to some 10 % of the incident intensity, although the exact amount depends upon zenith angle and upon water-vapour and ozone concentrations.

Solar radiation is also greatly modified by scattering processes, with the result that the stratosphere is irradiated by a second, diffuse stream of solar radiation, which may be regarded as coming from about 3 km. above the earth's surface with intensity equal to 40 % of the incident radiation. The spectral distribution of energy in the diffuse stream differs greatly from that of the incident stream, since the scattering may not be 'white' (e.g. the molecular scattering follows Rayleigh's λ^{-4} law), and since the double path through the troposphere means that absorption will already be very strong in the regions where the stratosphere can absorb. Thus the scattered radiation will not be of great energetic importance in the stratosphere, although it can scarcely be ignored.

In order to maintain an overall heat balance, the earth and atmosphere emit to space radiation which corresponds approximately to black-body radiation at 250° K. The majority of this radiation is contained in the wave-length range 4–60 μ, where the atmosphere has many intense absorption bands of water vapour, carbon dioxide and ozone. By Kirchhoff's law, therefore, the atmosphere can emit radiation similar in amount to that which it can absorb from this terrestrial stream of radiation. This means that the problem of low-temperature radiation in the earth's atmosphere is a transfer problem which is of such complication that only graphical methods of solution have been used up to now, and even these usually

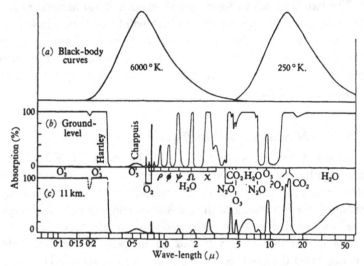

Fig. 53. (a) Black-body emission for 6000 and 250° K. (b) Absorptivity of the atmosphere from the ground level to outer space for diffuse radiation or for a solar beam at 50° zenith angle. (c) The same for the atmosphere from the temperate tropopause to outer space. It has been assumed that the atmosphere contains 3 mm. of ozone at S.T.P. all above 11 km.; that it contains 2 g.cm.$^{-2}$ of water vapour above ground level and 10^{-3} g.cm.$^{-2}$ above 11 km. The small window (dotted) at 0·2 μ and the ozone band marked (?) have not yet been observed at the levels concerned. (After Goody and Robinson, 1951.)

contain many unwarranted simplifications. The problems are rendered all the more difficult by the fact that all the important absorption bands are vibration-rotation bands with discrete line structure.

Fig. 53 gives a general idea of the magnitudes of the energies involved in the various absorption processes. Diagram (a) shows black-body curves representing the solar and terrestrial fluxes. These two curves are drawn with equal areas, since, averaged over all angles, all time, and the whole surface of the earth, the two fluxes must balance. Diagrams (b) and (c) give the absorptivity of the whole atmosphere and of the stratosphere respectively. Thus the product of (c) and the solar curve in (a) gives the solar energy absorbed by the stratosphere, while the product with the second curve in (a) gives the energy emitted by the stratosphere to space; these are not equal to each other, since radiation is also absorbed from the troposphere and the earth's surface.

The bands shown in fig. 53 are all smoothed out to obscure any fine structure, the importance of which will be discussed in §6.21.

A point of general importance, which is obvious from fig. 53, is that three minor constituents only are responsible for the main entries in the stratospheric heat budget, viz. water vapour, carbon dioxide and ozone.

6.12. *The validity of Kirchhoff's law*

In the following sections use will be made of Kirchhoff's radiation law by equating the energy emitted by an element of the atmosphere to the product of the absorptivity with the black-body radiation corresponding to the gas-kinetic temperature. Although this law was deduced for cavity radiation, it has been shown by Milne (1930) that it is valid for an unenclosed radiation field provided that the gas-kinetic temperature can be specified (i.e. that there is a Maxwellian distribution of velocities), and that the energy levels responsible for the absorption and emission are in thermal equilibrium (i.e. are populated according to Boltzmann's law).

According to Spitzer (1949), Maxwell's law of distribution of velocities is obeyed throughout the stratosphere. The condition for thermal equilibrium between the energy levels has been reduced by Milne (1930) to the condition that the rate of de-excitation from the excited states by inelastic collisions should be greater than the rate of de-excitation by spontaneous and radiation-induced transitions. Comparing vibrational and rotational transitions it can be shown that the excited rotational levels have longer lifetimes than the excited vibrational levels and that the efficiency of collisional de-excitation is very much higher (Zener, 1931 a). It is therefore the population of the vibrational levels which must be critically examined.

In Table XVI are given the natural lifetimes for the excited states responsible for three typical atmospheric absorptions. The lifetimes under atmospheric conditions will be slightly shorter owing to additional, induced transitions, but this effect is small at atmospheric temperatures. It can be seen that the rate of de-excitation by spontaneous transitions is of the order of 10 per sec. per molecule.

The efficiency of de-excitation from vibrational states by collisions is very small. According to Zener (1931 *a* and *b*) it may

TABLE XVI. *Lifetimes of excited vibrational states*

Gas	Band centre (μ)	Band intensity (c./sec. per g.cm.$^{-2}$)	Lifetime (sec.)
H_2O	6·3*	$8·1 \times 10^{15}$	6×10^{-2}
N_2O	7·8†	$4·4 \times 10^{15}$	9×10^{-2}
CO_2	15‡	$3·0 \times 10^{15}$	4×10^{-1}

* Cowling (1950).   † Goody and Wormell (1951).   ‡ Kaplan (1950).

The band intensities have all been multiplied by the vibrational partition function (see Moelwyn-Hughes, 1940) corresponding to the temperature of the observation in order to obtain the value appropriate to all the molecules being in the ground vibrational state.

lie between 10^{-4} and 10^{-5} for resonant collisions between diatomic molecules, while the lack of resonance may be responsible for a further factor of 10^{-1} to 10^{-2}. These figures are based upon approximate quantum-mechanical arguments, but they roughly confirm values deduced from measurements of the velocity of sound at high frequencies. Henry (1931) finds an efficiency of 5×10^{-6} for collisions between oxygen molecules in this way, although recent measurements show it to be greater for polyatomic molecules.

Using this figure together with the data in Table VI (Chapter III) we find that the rate of collisional de-excitation is of the order of 400 per molecule per sec. at 32 km., and 7 per molecule per sec. at 62 km. Comparing with the rate of spontaneous de-excitation we see that Kirchhoff's law is certainly valid up to 50 km., but may fail above this level.

The numerical work described in § 6.3 only applies to levels below 55 km., and the use of Kirchhoff's law is unlikely to cause any considerable errors. Although it may not be strictly valid, its use may not involve very large errors above this level, nevertheless it must be hoped that in the future it will be possible to work with the more exact equation of transfer of radiation due to Eddington (1929), although this will require information about inelastic collisions which is not at present available.

6.13. *The integral equations*

In the following treatment we will consider a plane stratified atmosphere, i.e. we neglect the earth's curvature, horizontal variations in the terrain, variable cloud cover, etc. Scattered solar radiation will also be neglected because it has already traversed a very long path and will not yield much more energy in the regions of atmospheric absorption. Low-temperature terrestrial radiation will be assumed not to be scattered, and the lower boundary condition will be idealized to the extent of replacing the earth's surface by a black body at the same temperature. If cloud has to be taken into account it will be assumed that it is sufficient to raise the effective level of the earth's surface. Finally, the solar radiation will be assumed to be parallel. All these simplifications and assumptions have been accepted uncritically up to the present time, and indeed a close examination can safely be postponed until the simplified problem has been treated effectively.

The following symbols will be required:

z, z', z'' = vertical co-ordinate,

θ = temperature,

ν = frequency

$\frac{1}{\pi} B_\nu(\theta) \Delta\nu$ = energy cm.$^{-2}$sec.$^{-1}$ per unit solid angle emitted in the frequency range ν to $\nu + d\nu$ by a black body at temperature θ,

k_ν = absorption coefficient per g.cm.$^{-2}$ at frequency ν,

ρ = density of absorbing gas,

ρ_a = density of the atmosphere,

c_p = specific heat of air at constant pressure,

ψ = angle from the zenith,

$S_\nu \Delta\nu$ = solar energy cm.$^{-2}$sec.$^{-1}$ incident on the atmosphere in the frequency range ν to $\nu + d\nu$.

In deriving the integral equations we shall make use of Kirchhoff's law and of Lambert's exponential law of absorption, which is invariably correct for strictly monochromatic radiation. Radiation fluxes will be counted positive when they transfer energy upwards.

The solar flux outside the atmosphere per unit area of the earth's surface is $S_\nu \Delta \nu \cos \psi_s$, where ψ_s is the solar zenith angle, and the flux reaching a level z will be the product of this and the transmission of the atmosphere above z;

$$\phi_\nu^S(z)\Delta \nu = - S_\nu \Delta \nu \cos \psi_s \exp \left[- \int_z^\infty k_\nu(z')\rho(z')\sec \psi_s dz' \right].$$

To evaluate the flux from the earth's surface reaching z, we need to integrate over all angles. The emission per cm.[2] of the earth's surface ($z = 0$) per unit solid angle at zenith angle ψ is

$$\frac{1}{\pi}B_\nu(0)\Delta \nu \cos \psi,$$

while the solid angle embraced by the angles ψ and $\psi + d\psi$ is

$$2\pi \sin \psi \, d\psi.$$

The radiation reaching z is therefore given by

$$2B_\nu(0)\Delta \nu \cos \psi \sin \psi \, d\psi \, \exp \left[- \int_0^z k_\nu(z')\rho(z')\sec \psi \, dz' \right].$$

The integration over ψ from 0 to $\frac{1}{2}\pi$ may be performed using the exponential integral

$$\mathrm{Ei}_n(x) = \int_1^\infty e^{-x\eta}\frac{d\eta}{\eta^n},$$

by means of the substitution $\eta = \sec \psi$, to give

$$\phi_\nu^E(z)\Delta \nu = 2B_\nu(0)\Delta \nu \, \mathrm{Ei}_3 \left[\int_0^z k_\nu(z')\rho(z') \, dz' \right].$$

The final contribution to the flux at z is from the atmosphere above and below z. Consider a level z' below z. The absorptivity of a layer dz' at zenith angle ψ is $\rho(z')k_\nu(z')\sec \psi dz'$. The black-body emission at an angle ψ has already been given, and the radiation emitted in the solid angle embraced by ψ to $\psi + d\psi$ is therefore

$$2B_\nu(z')\Delta \nu \rho(z')k_\nu(z') \, dz' \sin \psi \, d\psi,$$

while the part which reaches z will be

$$2B_\nu(z')\Delta \nu \rho(z')k_\nu(z') \, dz' \sin \psi \, d\psi \, \exp \left[- \int_{z'}^z k_\nu(z'')\rho(z'')\sec \psi \, dz'' \right].$$

The integration over ψ can be performed with the help of the second exponential integral, giving for the radiation from all levels which reaches z

$$\phi_\nu^B(z)\,\Delta\nu = \int_0^z 2B_\nu(z')\,\Delta\nu\rho(z')\,k_\nu(z')\,\mathrm{Ei}_2\left[\int_{z'}^z \rho(z'')\,k_\nu(z'')\,dz''\right]dz'$$

$$= \int_0^z 2B_\nu(z')\,\Delta\nu\frac{\partial}{\partial z'}\,\mathrm{Ei}_3\left[\int_{z'}^z \rho(z'')\,k_\nu(z'')\,dz''\right]dz'.$$

Similarly, from all levels above z there is a flux of radiation

$$\phi_\nu^A(z)\,\Delta\nu = \int_z^\infty 2B_\nu(z')\,\Delta\nu\frac{\partial}{\partial z'}\,\mathrm{Ei}_3\left[\int_z^{z'} \rho(z'')\,k_\nu(z'')\,dz''\right]dz'.$$

To obtain the total flux at z an integration has to be performed over all frequencies,

$$\phi(z) = \int_0^\infty \left[\phi_\nu^S(z) + \phi_\nu^E(z) + \phi_\nu^A(z) + \phi_\nu^B(z)\right]d\nu,$$

and the rate of heating per unit volume is

$$\frac{\partial Q}{\partial t} = \rho_a c_p \frac{\partial\theta}{\partial t} = -\frac{\partial\phi(z)}{\partial z}.$$

The obstacles in the way of a numerical solution of these equations are formidable. We may list the three worst as follows:

(a) Since bands to some extent overlap, the factor ρk_ν is a sum over at least three components, each of which has a different spatial distribution.

(b) The water-vapour and ozone concentrations and the temperature at any level are very variable, and, unless all the integrations are to be repeated upon every occasion, it is necessary to obtain a solution into which arbitrary values of these variables may be put (i.e. what is commonly called a 'radiation diagram').

(c) Three successive integrations must be performed, the significance of which is best appreciated if we consider the number of columns required to set out a numerical calculation. Each spatial integration might be handled by means of thirty intervals, and two consecutive, dependent, integrations would involve 10^3 columns in a computation. The labour involved in the frequency integration depends upon the frequency interval over which the radiation may be termed 'monochromatic'. Vibration-rotation bands are so complicated that at least 10^4 intervals are required, and since the integration is dependent upon the previous spatial integrations, this demands a computation involving 10^7 columns. Even if it were possible or desirable to code this problem, there are few electronic digital computers which could solve it. Some steps towards the simplification of the problem by means of approximations are therefore unavoidable. It should be possible to devise a sufficiently exact method whereby each atmospheric component can be treated separately, each with its own radiation diagram. It is also possible that the interaction between the two spatial integrations can be taken into account by means of approximations. The most formidable problem is the frequency integration and its dependence upon the spatial integrations. It is absolutely necessary to devise a method of performing the frequency integration once and for all, and to do so requires a careful study of the physics of molecular absorption.

6.2. Absorption by molecular bands

6.21. *The nature of a vibration-rotation band*

Strong vibration-rotation bands owe their existence to a transition from the ground state to one of the first vibrational states, and their fine structure to simultaneous transitions between rotational states. The selection rules for a rigid rotator allow transitions for which the rotational quantum number changes by 0 or ± 1 or alternatively by ± 1 only, depending upon the symmetry conditions. For no change in the rotational quantum number a narrow Q branch is observed at the band centre, while for a change of -1 a long-wave P branch is formed, and for a change of $+1$ a short-wave R branch. Since the rotational energy is not a linear function of the quantum number, the P and R branches are spread out and formed of discrete lines. For example, in the case of a linear molecule without a centre of symmetry whose moment of inertia (I) does not depend upon the vibrational state, the lines are spaced at equal intervals

$$\Delta \nu = 2B = \frac{h}{8\pi^2 cI},$$

where B is the rotational constant, and h is Planck's constant.

Nitrous oxide falls into this class of molecule, and has a rotational constant of 0·418 cm.$^{-1}$.† The fundamental band at 7·8 μ has no Q branch, and the broad P and R branches, with a fine structure spacing of about 0·84 cm.$^{-1}$, are shown in a spectrum of Plyler and Barker (1931) (see fig. 54).

The carbon dioxide molecule is also linear, with a rotational constant of 0·390 cm.$^{-1}$, but, being symmetrical, alternate lines are missing, and the mean line spacing in the P and R branches is approximately 1·56 cm.$^{-1}$. The important fundamental band centred at 15 μ also has a sharp and intense Q branch.

The problems become very much more complicated for asymmetric molecules such as ozone and water vapour. Water vapour

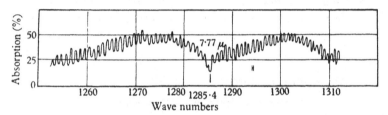

Fig. 54. Fine structure of the 7·8 μ fundamental of nitrous oxide.
(After Plyler and Barker, 1931.)

has three rotational constants, viz. 27·8, 14·5 and 9·3 cm.$^{-1}$, while for ozone the values are believed to be 5·00, 0·414 and 0·383 cm.$^{-1}$. Nevertheless, the quantum theory of such molecules has been worked out (Cross, Hainer and King, 1944), and extremely good agreement with observation has been obtained for water vapour. Fig. 55 shows a spectrum of the water-vapour rotation band (for which there is no change in the vibrational state and which is therefore 'centred' on zero frequency) together with some early calculations of line positions and intensities.

More precise calculations than those shown in fig. 55 have been performed for the rotation band and also for the 6·3 μ band of water vapour by Yamamoto and Onishi (1949). Similar calculations have not yet been made for ozone, about whose molecular structure there is still some doubt.

† cm.$^{-1}$ is the unit of wave number (equal to 10⁴ divided by the wave-length in μ).

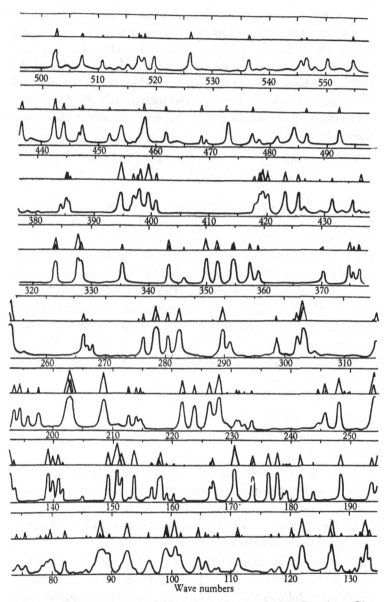

Fig. 55. The water-vapour rotation band. (After Randall, Dennison, Ginsberg and Weber, 1937.) N.B. Different amounts of water vapour were in the path for the different ranges; the strongest lines by far are in the region of 200 cm.⁻¹. The triangles represent calculated positions and intensities of water-vapour lines.

We can see therefore that the positions of all the lines in the atmospheric absorption spectrum can be specified in terms of a few simple molecular constants. The same constants, together with the temperature, also specify the relative intensities of the lines in one band, since these only depend on the population and the statistical weights of the various rotational levels of the molecule. For example, for linear molecules, the line intensity in absorption is given by

$$k \propto \frac{hcB}{k\theta}(J'+J''+1)\exp\left[-\frac{BJ''(J''+1)\,hc}{k\theta}\right],$$

where J' and J'' are the quantum numbers of the upper and lower rotational states respectively. This relation gives the characteristic intensity maxima of P and R branches exhibited in fig. 54.

Where the molecular structure is well established therefore, it is only necessary to know the sum of all line intensities in a band to discover the intensity of each line. This quantity (the band intensity) is difficult to calculate for vibrational transitions, but it may be measured experimentally. Some band intensities have already been given in Table XVI. In addition to those shown there, the work of Cowling (1950) and Yamamoto and Onishi (1949) gives data for the water-vapour rotation band.

6.22. *Line shapes*

The shape of an individual line in a vibration-rotation band depends upon the following factors:

(*a*) the natural lifetime of the excited vibrational state,

(*b*) Doppler effects from molecular thermal motion,

(*c*) collisions which shorten the natural lifetime,

(*d*) the mutual distortion of molecular structures and consequently of the molecular energy levels.

In §6.12 we established that natural lifetimes are far longer than the time between collisions at all levels in the stratosphere. From this it follows that (*a*) may be neglected in comparison with (*c*). In practice we may also neglect (*d*) for pressures less than 1 atmosphere (Goody and Wormell, 1951).

The effect of (*c*) is to replace the natural lifetime in the Lorentz or 'dispersion' expression for line shape† by the artificial lifetime caused by molecular collisions. The absorption coefficient is therefore given by

$$k_\nu = \frac{k}{\pi} \frac{\alpha_L}{(\nu - \nu_0)^2 + \alpha_L^2}, \quad \alpha_L = \frac{1}{2\pi\tau},$$

where α_L is the half-line width (in contrast to the spectroscopic line width $2\alpha_L$), ν_0 is the frequency of the line centre, and τ is the time between collisions. According to kinetic theory the half-line width is

$$\alpha_L = \frac{1}{4\pi} \sum_i N_i (D + D_i)^2 \left[2\pi k\theta \left(\frac{1}{m} + \frac{1}{m_i} \right) \right]^{\frac{1}{2}},$$

where N_i = the number of perturbing molecules of the ith class,

$D + D_i$ = the sum of the optical collision diameters for an absorbing and a perturbing molecule,

m = the mass of an absorbing molecule,

m_i = the mass of a perturbing molecule.

If the absorbing molecules are dilute and the perturbing molecules are mixed in the same proportions throughout the atmosphere, N_i is proportional to the total atmospheric pressure. Since pressure varies so much over the stratosphere this is a relation of great importance.

Complications with the above relation for half-line width are that collision diameters depend upon the rotational quantum number and also upon the temperature. These factors are of small importance compared with the large pressure effect, and therefore in Table XVII only one value of the Lorentz width is assigned to each atmospheric constituent. The Lorentz widths for carbon

† In an attempt to avoid complications which may eventually turn out to be irrelevant we will not use the exact expression for the shape of a line broadened by elastic collisions (van Vleck and Weisskopf, 1945) which differs from the Lorentz expression in the far wings, and further we will neglect the fact that distortion of the molecular structure must take place during a collision, which again makes the Lorentz expression inexact in the wings. Plass and Warner (1952) show that for the single-line model (§6.23) the shape in the wings is important for large absorptions. In the body of an absorption band, however, each individual line is surrounded by other intense lines and the importance of the exact shape in the wings is greatly reduced. There is as yet no experimental or theoretical evidence that the Lorentz expression is not an adequate approximation for the main part of infra-red absorption bands. The residual absorption well away from a band centre will of course depend critically upon the shape in the line wings, but such absorptions are probably not of great consequence in stratospheric problems.

dioxide and ozone have not been measured, but their collision diameters may be estimated from the known collision diameters for momentum interchange.

TABLE XVII. *Lorentz and Doppler half-line widths for the main atmospheric absorption bands*

Gas	Band centre (μ)	α_D at 300° K. (cm.$^{-1}$ × 10^2)	α_L at S.T.P. (cm.$^{-1}$)	p (mm. Hg)	h (km.)
H_2O	2·7	6·4	0·11[*]	44	17
	6·3	2·8		19	23
	20 } rotation	0·88		6	32
	40 } band	0·44		3	37
CO_2	4·3	2·6	0·15	13	26
	15	0·75		4	34
N_2O	7·8	1·5	0·16[†]	7	30
O_3	4·7	2·3	0·16	11	27
	9·6	1·1		5	33
	14·1	0·76		4	34
CH_4	3·3	5·6	0·18[‡]	24	22
	7·7	2·4		10	27

[*] Adel (1947a), Becker and Autler (1946).   [†] Goody and Wormell (1951).
[‡] Goldberg (1951).

Table XVII also shows Doppler half-line widths and, in the final two columns, the pressure and height in the atmosphere at which Doppler and Lorentz widths become equal to each other.

Doppler widths may be calculated from Maxwell's law, which gives for the proportion of molecules with velocities between u and $u + du$ in the line of sight,

$$p(u)\,du = \left(\frac{m}{2\pi k\theta}\right)^{\frac{1}{2}} \exp\left[-\frac{mu^2}{2k\theta}\right] du.$$

If there is a narrow absorption line at ν_0, Doppler effects will make it appear to be at a frequency $\nu_0 \pm \dfrac{u}{c}\nu_0$ leading to a line shape

$$k_\nu = \frac{k}{\alpha_D \sqrt{\pi}} \exp -\left[\frac{\nu - \nu_0}{\alpha_D}\right]^2,$$

where α_D (the Doppler half-line width) is given by

$$\alpha_D = \frac{\nu_0}{c}\left(\frac{2k\theta}{m}\right)^{\frac{1}{2}}.$$

A comparison between the Doppler and Lorentz shapes is shown in fig. 56. The Doppler line is more compact than the Lorentz line, having less intensity in the wings and more at the centre.

A pure Doppler line cannot exist in practice, since the effect of molecular motions is superimposed upon the effect of a finite life-time. From the Maxwell distribution, the combined effect of these two factors must lead to the line shape

$$k_\nu = k \int_{-\infty}^{+\infty} p(u) \frac{\alpha_L}{\pi} \frac{du}{\left(\nu - \nu_0 + \dfrac{u}{c}\nu_0\right)^2 + \alpha_L^2}.$$

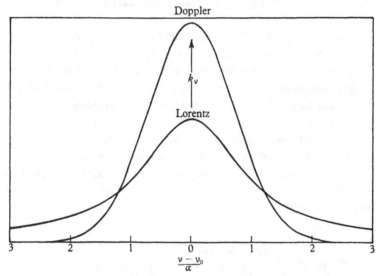

Fig. 56. Doppler and Lorentz line shapes for similar
intensities and half-line widths.

For $a = 2\alpha_L/\alpha_D > 1$, the integral can for practical purposes be replaced by the Lorentz formula. For $a \leqslant 1$, a numerical solution has been obtained by van der Held (1931), who showed that the equation leads to a hybrid line with a centre similar to a Doppler line and with the wings of a Lorentz line.

6.23. Absorption by a single line

Since B_ν is a slowly varying function of frequency as compared with the absorption coefficient, it is normally possible to subdivide the frequency integration into ranges over which B_ν may be considered to be constant but which contain many absorption lines. Taking as an example the flux of radiation transmitted from the earth's surface, we may write

$$\phi^E(z) = \int_0^\infty 2B_\nu(0)\,\mathrm{Ei}_3\left[\int_0^z k_\nu(z')\rho(z')\,dz'\right]d\nu$$

$$= \sum_{\text{all } r} 2B_r(0)\int_r \mathrm{Ei}_3\left[\int_0^z k_\nu(z')\rho(z')\,dz'\right]d\nu,$$

where it is usually possible to arrange that the integral is a slowly varying function of the position of the range in the spectrum. The quantity multiplying B_r is commonly called a 'transmission function'.

Roberts (1930) has shown that the function $2\mathrm{Ei}_3(x)$ can be very closely approximated by the function $\exp(-1\cdot5x)$, from which we may conclude that the equations of transfer for diffuse radiation may be approximated to the equations for parallel radiation simply by increasing the absorbing matter by 50 %.† At this stage therefore we need only discuss the equations for parallel radiation.

Let us first assume that lines are so far apart that overlap between them may be neglected. This clearly implies that the mean transmission over several lines is very close to unity, and therefore this model cannot often be used in practice. The transmission of m g.cm.$^{-2}$ along a path at constant pressure after averaging over one line is

$$T^*(m) = \frac{\int_{-\frac12\delta}^{+\frac12\delta} \exp[-k_\nu m]\,d\nu}{\int_{-\frac12\delta}^{+\frac12\delta} d\nu} = 1 - \frac{1}{\delta}\int_{-\frac12\delta}^{+\frac12\delta}\{1 - \exp[-k_\nu m]\}\,d\nu,$$

where δ is the mean spacing between lines. The assumption that lines do not overlap allows the limits in the last integral to be replaced by $\pm\infty$, and, substituting the Lorentz shape, the integration may be performed to give

$$T^*(m) = 1 - 2\pi y\gamma\, e^{-\gamma}\,[J_0(i\gamma) - iJ_1(i\gamma)],$$

where $y = \alpha_L/\delta$, $\gamma = km/2\pi\alpha_L$, and the J's are Bessel functions of the first kind. The two non-dimensional coefficients which have been introduced here will be used in the following sections as well.

† More recent treatments of the regular model (§6.25) give a value of 66 % (Elsasser, 1942; Kaplan, 1952).

The above expression represents a simple transition from a linear law of absorption ($A^* = 1 - T^*$) to a square-root law, as m increases.

For $\gamma \ll 1$,
$$T^*(m) \to 1 - 2\pi y\gamma = 1 - \frac{km}{\delta},$$

while for $\gamma > 3$,
$$T^*(m) \to 1 - 2y\sqrt{(2\pi\gamma)} = 1 - \frac{2\sqrt{(k\alpha_L m)}}{\delta}.$$

The physical significance of the linear law of absorption is that the absorption is so small that all exponentials may be reduced to their

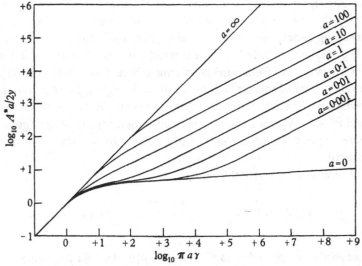

Fig. 57. Absorption by mixed Doppler and Lorentz lines. $a = 2\alpha_L/\alpha_D$. The ordinate and abscissa are chosen to give a convenient diagram. (After van der Held, 1931.)

first term in m; the linear law will therefore exist for any line shape. The square-root law, on the other hand, is characteristic of the effect of the wings of a Lorentz line after the centre has been fully absorbed.

For mixed Doppler and Lorentz line shapes, the required integration has been performed numerically by van der Held (1931), and the results are shown in fig. 57.

The curves for $a > 1$ in fig. 57 are given by the Bessel function relation for Lorentz lines, and they show the simple transition from a linear to a square-root law as γ increases. For pure Doppler broadening ($a = 0$), absorption increases only very slowly after the linear law region, because the intensity of a Doppler line falls off so rapidly in the wings. Only for $a < 1$ do Doppler effects influence the absorption, and for all values of a they may be neglected if the absorption is large enough; this is simply a restatement of the fact that in its far wings a mixed line always has the Lorentz shape.

So far we have only considered absorbing paths at constant pressure, which begs the question of the interaction between the integrations in the flux equations. A ray passing downwards through an atmosphere wherein the lines have the Lorentz shape will pass through layers with continually increasing line width, and the resulting superposition of lines will never itself have the Lorentz shape. One case only has been worked out exactly, and that is for the transmission from outside an atmosphere down to an arbitrary level for a gas of constant mixing ratio at all heights. The expression given here was first derived by Pedersen (1942) and later by Strong and Plass (1950) for an isothermal atmosphere; the treatment given here applies, however, to an atmosphere of arbitrary thermal structure.

We wish to evaluate the expression

$$A^*(z)\,\delta = \int_0^\infty d\nu \left\{ 1 - \exp\left[-\int_z^\infty \rho(z')\, k_\nu(z')\, dz' \right] \right\}.$$

Let $m(z') = \int_z^\infty \rho(z')\, dz'$, so that $dm = -\rho(z')\, dz'$. By definition we can write

$$m(z) = \frac{2\pi\gamma}{k}\, \alpha_L(z),$$

and from the hydrostatic equation we can show that γ is not a function of height. From the hydrostatic equation

$$-dp(z) \propto \rho_a(z)\, dz,$$

and since the mixing ratio is constant $\rho(z) \propto \rho_a(z)$ and $dm(z) \propto dp(z)$. Since $p(\infty) = m(\infty) = 0$, m and p are proportional, and from the Lorentz formula p and α_L are also proportional, so that γ is not a function of height.

For γ constant the spatial integration may be performed for Lorentz lines, and we obtain the result

$$A^*(z) = \frac{1}{\delta}\int_0^\infty dv\left\{1 - \exp\int_{\alpha_L(z)}^0 \frac{2\gamma\alpha_L d\alpha_L}{(v-v_0)^2+\alpha_L^2}\right\}$$

$$= \frac{1}{\delta}\int_0^\infty dv\left\{1 - \left[\frac{(v-v_0)^2}{(v-v_0)^2+\alpha_L^2(z)}\right]^\gamma\right\} = 2y(z)\,\pi^{\frac{1}{2}}\frac{\Gamma(\gamma+\frac{1}{2})}{\Gamma(\gamma)}.$$

This simple expression is unfortunately of little practical application as it stands, since, as has already been pointed out, a single-line approximation to real band absorption can only be valid if the absorption is very small. In fact, the above expression tends to infinity for large γ, while in practice the absorption can never exceed unity. Nevertheless, the expression has been used by Strong and Plass for large absorptions when it must be very greatly in error.

6.24. *Absorption by a disordered band*

The case of absorption by a disordered band has been treated by Goody (1952), who showed that, assuming no correlation between line intensities and positions, and also that all combinations of line positions are equally probable, then the expression for the mean transmission over a large number of lines can be integrated in the form

$$T^*(\alpha, m) = \exp\left[-\frac{1}{\delta}\int_{-\infty}^{+\infty} dv\int_0^\infty dk\, P(k)\{1 - \exp[-mkS(v,\alpha)]\}\right],$$

where $dk\,P(k)$ is the probability that a single line has an intensity in the range k to $k+dk$, $S(v,\alpha) = k_v/k$ is a line shape parameter, and δ is the mean spacing of the lines. The expression can be applied either to Lorentz or Doppler broadening or to absorption along a pressure gradient by substituting different expressions for $S(v,\alpha)$, and it immensely simplifies the problem of treating a disordered band by reducing its solution to that of an integral over one line only.

A glance at the fine structure of the water-vapour bands illustrated in fig. 55 suggests that this model might be used to represent the absorption by this important gas. That this is so can be shown by substituting the Lorentz shape and comparing the result with a series of elaborate calculations by Cowling (1950) using the full quantum-mechanical data upon line positions and intensities under a few definite conditions of constant pressure. Before the comparison can be made, an expression is required for $P(k)$, but it can be shown that the precise choice is not important and that the distribution

$$P(k) = \frac{1}{\sigma} \exp\left[-\frac{k}{\sigma} \right]$$

is quite adequate, where σ is the average line intensity. This distribution, together with the Lorentz line shape, leads to the expression

$$T^*(\alpha_L, m) = \exp\left[-\frac{m\sigma\alpha_L}{\delta\left(\alpha_L^2 + \frac{m\sigma\alpha_L}{\pi}\right)^{\frac{1}{2}}} \right] = \exp\left[-\frac{2\pi y\gamma}{(1 + 2\gamma)^{\frac{1}{2}}} \right],$$

which agrees very closely with Cowling's computations.

The applicability of the model to water vapour is therefore established, and there is some reason to suppose that it may also apply to ozone. Although this leads to a very great simplification, the problem is still complex if it is necessary to substitute a different value for the line shape for each distribution of absorbing matter with height. This question has been investigated by Curtis (1952), who has been able to show that, to a very high degree of accuracy, it is only necessary to make the line width, rather than the whole line shape, dependent upon the distribution. He has demonstrated numerically that in all important cases it is possible to consider all the absorbing matter to be at one pressure, given by

$$\bar{p} = \frac{\int p(z)\rho(z)\,dz}{\int \rho(z)\,dz}.$$

It is possible to show that this approximation is asymptotically correct under three different circumstances, viz. in the linear absorption region, for $\gamma > 10$ and for $y > 1$. This means that it is correct in the limits γ small and large and independently in the limits y small and large, and Curtis could show that it also provides an excellent interpolation formula for intermediate values of γ and y.

It is appropriate to mention that in most meteorological literature a different procedure is followed, often referred to as a 'scaling approximation', where it is assumed that the line shape under a pressure gradient can be represented by a Lorentz shape at any desired pressure by reducing the density of absorbing gas in each layer by a factor $(p(z)/p)^{\frac{1}{2}}$, where p is the arbitrarily selected pressure. It can be shown that this method is seriously in error in the three limiting cases where Curtis's method is correct and, moreover, that it is never exact.

6.25. *Absorption by a regular band*

The formalization of the last section is adequate for water vapour and perhaps for ozone, but it will clearly not suffice for carbon dioxide bands, consisting of equally spaced lines with slowly varying intensities. An obvious model for bands of this type is an infinite array of equally spaced lines of equal intensity.

If k is the line intensity and δ the line spacing, then the transmission at a frequency ν distant from the nearest line is clearly (Elsasser, 1942)

$$T_\nu = \exp - \sum_{-\infty}^{+\infty} m k_{\nu-n\delta},$$

which, for the Lorentz shape, can be shown to be equivalent to

$$T_\nu = \exp\left[-\frac{2\pi y \gamma \sinh 2\pi y}{\cosh 2\pi y - \cos 2\pi x} \right],$$

where $x = \nu/\delta$.

The mean value of the transmission can be found by means of an integration from $x = -\frac{1}{2}$ to $+\frac{1}{2}$, and hence

$$T^*(y,\gamma) = \int_{-\frac{1}{2}}^{+\frac{1}{2}} \exp\left[-2\pi \gamma y \frac{\sinh 2\pi y}{\cosh 2\pi y - \cos 2\pi x} \right] dx.$$

This useful function has not yet been tabulated, although Elsasser has derived three approximations:

(*a*) for $y \gg 1$† $T^*(y,\gamma) \to \exp[-2\pi y \gamma]$,

(*b*) for $y \ll 1$ and T^* near unity
$$T^*(y,\gamma) \to 1 - 2\pi \gamma y\, e^{-\gamma}\, [J_0(i\gamma) - iJ_1(i\gamma)],$$

(*c*) for $\gamma \gg 1$ $T^* \to 1 - \mathrm{erf}(\pi y[2\gamma]^{\frac{1}{2}})$.

† The precise conditions of validity for these and the following three approximations are very complicated; the conditions stated only give a rough guide.

In (a) the line structure has been smoothed out and Beer's law is obeyed, while (b) refers to conditions under which lines can be treated as not overlapping. Unfortunately, it may often happen that none of these approximations is admissible in the troposphere.

Pedersen's calculation for a mixed atmosphere (§6.23) gives, for the shape of a single line, the expression

$$k_\nu = \frac{k}{2\pi\alpha_L} \log_e \frac{(\nu-\nu_0)^2}{(\nu-\nu_0)^2 + \alpha_L^2(z)},$$

which leads to

$$T^*(y,\gamma) = \int_{-\frac{1}{2}}^{+\frac{1}{2}} \left\{ \prod_{-\infty}^{+\infty} \frac{(x-n)^2}{(x-n)^2+y^2} \right\}^\gamma dx$$

$$= \int_{-\frac{1}{2}}^{+\frac{1}{2}} \left\{ \frac{1-\cos 2\pi x}{\cosh 2\pi y - \cos 2\pi x} \right\}^\gamma dx.$$

This function also has three approximate forms:

(a) for $y \gg 1$ $T^* \to \exp[-2\pi y\gamma]$,

(b) for $y \ll 1$ and T^* near unity

$$T^* \to 1 - 2y\pi^{\frac{1}{2}} \frac{\Gamma(\gamma+\frac{1}{2})}{\Gamma(\gamma)},$$

(c) for $\gamma \gg 1$ $T^* \to 1 - \mathrm{erf}(\pi y\gamma^{\frac{1}{2}})$.

In this case the full transmission function has been tabulated for use with the experiments upon nitrous oxide described in §3.51, and a graphical representation is shown in fig. 58.

The limits of validity of the three approximations are illustrated in fig. 58. In the case of the experiments described in §3.51 none of the approximations was adequate, although for carbon dioxide in the stratosphere we may estimate from Table XVII and the known line spacing that $-\log_{10} y > 1\cdot 7$, and therefore that a combination of (b) and (c) will cover all important cases.

6.3. Numerical computations

6.31. *Solar absorption*

The work described in §6.2 has not yet been used to derive a rigorous radiation diagram which would enable the flux or its divergence to be calculated for any distribution of absorbing matter. In this and the next section we will describe two sets of computations limited to a few selected distributions and based

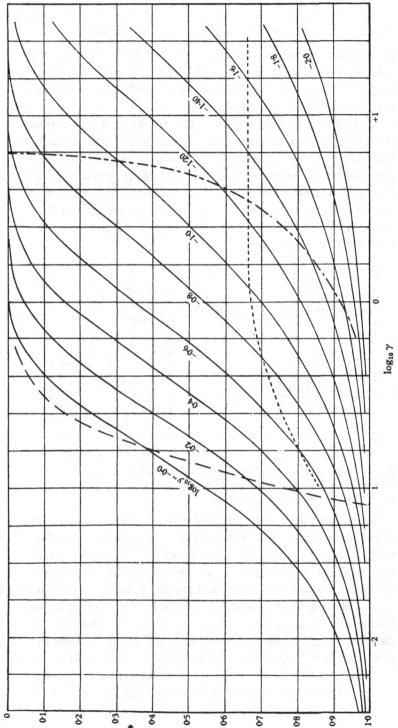

Fig. 58. $T^*(y, \gamma)$ for a gas with constant mixing ratio for radiation coming from outside the atmosphere. —— 1 % error limit for approximation (a); ······ 1 % error limit for approximation (b); —·—·— 1 % error limit for approximation (c).

upon treatments of gaseous absorption of dubious accuracy. We will not here discuss the methods used, although, despite all approximations, the computations undoubtedly give a valuable general guide to the magnitudes involved.

Karandikar (1946) has computed the absorption of energy from the solar beam assuming that the ozone distribution is that given by inversion effect measurements at Tromsö, that carbon dioxide forms 0·03 % of the atmosphere at all levels, and that water vapour is in diffusive equilibrium with four different total amounts of water vapour in the stratosphere (of which only the two smallest, viz. 5×10^{-3} and 5×10^{-2}, are compatible with the results of §3.2 for middle latitudes).

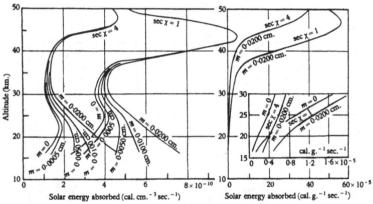

Fig. 59. The absorption of solar energy as a function of height in middle latitudes; m is the total water vapour in the stratosphere, and χ is the solar zenith angle. (After Karandikar, 1946.)

Fig. 59 gives the energy absorbed per unit mass as well as that absorbed per unit volume. The energy per unit mass abscissa, when multiplied by $3·6 \times 10^5$, gives the rate of rise of temperature in degrees per day in the absence of any cooling agencies, so that near 45 km. the rate of rise of temperature could be $200°$ K. day^{-1}. The heating at these levels is mainly attributable to absorption in the Hartley and Huggins bands of ozone. The energy absorbed by these bands falls off sharply between 45 and 30 km., because the absorption is very intense by the time the radiation reaches the lower of these levels. Below 20 km. the Chappuis bands in the visible spectrum can absorb more energy than the ultra-violet bands, and this leads to a secondary peak in the energy absorption at the level of maximum ozone concentration (25 km.).

The near infra-red water-vapour bands between 0·8 and 4·0 μ are responsible for the absorption of considerable energy in the lower stratosphere, the magnitude of which may be inferred from the curves for different amounts of water vapour. There is also a slight absorption of energy in the 2·7 and 4·3 μ carbon dioxide bands reaching a maximum of 0·8 × 10^{-10} cal.cm.$^{-3}$sec.$^{-1}$ at 12 km. for sec $\chi = 1$, and 0·2 × 10^{-10} cal.cm.$^{-3}$sec.$^{-1}$ at 19 km. for sec $\chi = 4$.

6.32. *Equilibrium temperatures*

The absorption of solar energy which takes place in the Hartley and Huggins bands near 50 km. is generally accepted as the reason behind the temperature maximum which exists in this region. The heat absorbed is dissipated by the emission of low-temperature radiation and by molecular and hydrodynamic processes, so that the average conditions in the stratosphere should be calculable by equating average absorption and average dissipation. If low-temperature radiation is taken to be the only form of energy dissipation, then the resulting equilibrium is termed 'radiative equilibrium'. Gowan (1947) has made a series of calculations of the radiative equilibrium temperature of the stratosphere up to 55 km., assuming that equilibrium is achieved at midday.

The discussion in the last chapter shows that turbulent transfer takes place at all times throughout the stratosphere, so that strict radiative equilibrium can never exist. Nevertheless, there is an important *a priori* justification for Gowan's assumption which shows that his results should be a good first approximation to the observed average temperatures. This justification lies in the fact that the absorption of solar radiation and the emission of low-temperature radiation are the source and sink of the partly irreversible heat engine responsible for the atmospheric motions; they are therefore the primary processes controlling the observed thermal structure of the atmosphere, and the hydrodynamic transfer can only have a modifying effect, although it may be a large one under some circumstances. In principle, however, it should be possible to compute possible atmospheric motions and their subsequent thermal effects from an initial state of radiative equilibrium, and in this respect Gowan's computations may be regarded as the first step.

Gowan divides the atmosphere into nine layers, each about 5 km. thick, from 11 to 55 km. The solar heating is computed by methods similar to those of Karandikar, for the midday sun in summer and winter at 50° N. The ozone distribution found by Götz and others (1934) is used, carbon dioxide is assumed to form 0·03 % of the atmosphere at all levels, while water vapour is assumed to have a constant mixing ratio with concentrations at the tropopause corresponding to 0, 10 and 40 % saturation. The absorption data used are rather crude, and, unlike Karandikar's computations, which mainly concern the pressure-independent electronic bands of ozone, the effect of pressure on the various molecular vibration bands is of very great importance. Gowan's method of including pressure effects might be approximately correct for Lorentz lines, but, as can be seen from Table XVII, the Lorentz shape will not hold above 40 km., and this method will therefore lead to emissivities which are too low and to equilibrium temperatures which are too high.

Whether or not the pressure effect upon water vapour is taken into account makes a considerable difference to the calculated equilibrium temperature, as can be seen from comparing fig. 60 (a) and (b). The same will apply to the ozone and carbon dioxide vibration-rotation bands, although, according to Gowan, carbon dioxide has little effect on the equilibrium above 30 km. Until the effect of pressure upon absorption can be properly included in these computations the results are clearly unreliable. However, the necessity for a region of high temperature from 50 to 60 km., as shown in fig. 60, will almost certainly remain even when the calculations have been improved.

To see whether the calculated high-temperature region will survive through the night, Gowan has computed the initial rate of change of temperature at various levels, assuming the sun to be cut off after equilibrium has been achieved. His results vary from 0·05 to 0·3° K. day^{-1} at 13 km., from 1·4 to 3·0° K. day^{-1} at 27·5 km., and from 39 to 96° K. day^{-1} at 47·5 km. These initial rates of change of temperature are considered by Gowan to be insufficient to destroy the high-temperature layer at night-time.

6.4. The thermal structure of the atmosphere

Gowan's computations assume mean values for the temperatures of the earth's surface and of the troposphere. While these assumptions probably have little effect upon the calculations above 30 km.,

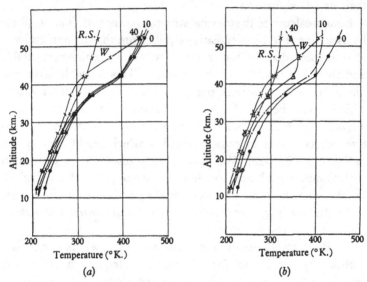

Fig. 60. Computed midday stratosphere temperatures for 50° N. (After Gowan, 1947.) (a) Pressure effects taken into account for all absorptions. (b) Pressure effects taken into account for all absorptions except those caused by water vapour.

Legend	Season	Relative humidity at 11 km. (%)	Solar colour temperature (° K.)
R.S.	Summer	10	4000
W.	Winter	10	6000
o	Summer	o	6000
10	Summer	10	6000
40	Summer	40	6000

the temperatures below this level, and in particular the temperature near to the tropopause, will depend very strongly upon them. In order to understand the nature of the transition at the tropopause, and such regular variations as are observed in the stratosphere below 20 km., we require a synthetic treatment which shows how the troposphere is built up and why the tropopause occurs where it does.

There have been many attempts at such a treatment (Gold, 1909; Emden, 1913; Hergesell, 1919; Milne, 1922; Dobson and others, 1946; Goody, 1949), and in this section we will discuss qualitatively their principal features.

Emden calculated the thermal structure of the earth's atmosphere by assuming it to be in radiative equilibrium throughout, and was able to show that the lowest layers would be extremely unstable and that they would therefore convect. This unstable layer was identified by Emden with the troposphere, and the layer above, which should be in radiative equilibrium, with the stratosphere. Thus Emden looked upon the tropopause as the point where radiative effects ceased to cause instability which might destroy the radiative equilibrium. We have already discussed in connexion with Gowan's work how this line of reasoning is perfectly logical. On this view the criterion governing the position of the tropopause is that the starting temperature gradient in the lower stratosphere should be just stable.

To illustrate Emden's important argument we may calculate with a very simple, but for this purpose adequate, model of the earth's atmosphere. The following assumptions will be made:

(a) The atmosphere has only one absorption coefficient for all wave-lengths from 4μ to ∞ ('grey absorption'), and it can absorb 90 % of the outward flux of radiation which originates at the earth's surface. This implies that

$$\exp\left[-\int_0^\infty k(z)\rho(z)\,dz\right] = 0.1.$$

(b) Water vapour is the only important absorbing material, and this is distributed exponentially with a scale height of 2 km., therefore

$$k(z)\rho(z) = k(0)\rho(0)\exp\left[-\frac{z}{2\times 10^5}\right],$$

and, from (a), $2\times 10^5 k(0)\rho(0) = 2.3$.

(c) From 0 to 4μ the atmosphere is transparent, so that all the solar radiation reaches the ground, where it is absorbed.

(d) The low-temperature radiation is confined to two anti-parallel streams.

Three relations can be shown to hold between the fluxes of terrestrial radiation at a point in space. They are direct consequences of the condition for radiative equilibrium, together with Kirchhoff's and Lambert's laws, and since they will be found in standard works on radiative transfer (Hopf, 1934) they will not be derived here.

$$\frac{\partial \chi(z)}{\partial z} = -k(z)\rho(z)\,\phi(z),$$

$$2B(z) = \chi(z),$$

$$\phi(z) = \text{const.},$$

where $\phi(z)$ is the integrated net upward flux, i.e. the difference between the upward and downward fluxes, $\chi(z)$ is the total radiation, i.e. the sum of the upward and downward fluxes, and, from Stefan's law, $B(z) = \sigma\theta^4(z)$. Clearly, for overall equilibrium of the planet, ϕ is equal to the incident flux of solar radiation. From the above equations we obtain the differential equation

$$\frac{\partial B(z)}{\partial z} = -\frac{k(z)\rho(z)}{2}\phi.$$

At the upper limit of the atmosphere there is no downward flux of long-wave radiation and therefore

$$\chi(\infty) = \phi \quad \text{and} \quad B(\infty) = \tfrac{1}{2}\phi.$$

At the earth's surface let the upward radiation flux be $B(G)$ corresponding to a temperature $\theta(G)$. Just above the earth's surface the net upward flux is ϕ, and therefore it follows that

$$\chi(0) = 2B(G) - \phi = 2B(0),$$

from the condition for equilibrium. There is therefore a negative discontinuity of temperature at the earth's surface, a characteristic situation at the boundary of a region in radiative equilibrium, whose importance will be emphasized later.

With these boundary conditions and the assumed distribution of absorbing matter, the differential equation may be solved with the result

$$\theta(z)^4 = \frac{\phi}{2\sigma}\left[1 + 2\cdot3\exp\left(-\frac{z}{2 \times 10^5}\right)\right],$$

and, for the ground temperature,

$$\theta^4(G) = \frac{4\cdot3\phi}{2\sigma}.$$

A value for the mean solar flux reaching ground level which is commonly used in meteorology is 0·271 cal.cm.$^{-2}$min.$^{-1}$, and substituting for ϕ we obtain the result shown in fig. 61. The ground temperature is 290° K., while the temperature of the lowest layer of the atmosphere is 271° K., giving an unstable discontinuity of 19° K. It is interesting to compare this value for the ground temperature with that which would have been obtained if there had been no atmosphere, which is 239° K. The shielding effect of the

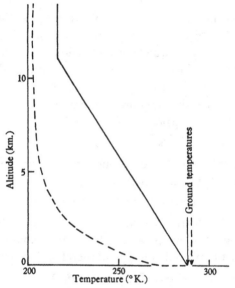

Fig. 61. Calculated and observed average temperature structure of the lower atmosphere. —— observed; --- calculated.

atmosphere is therefore responsible for raising the temperature by 51° K. The temperature at the upper limit of the atmosphere is 202° K., and the temperature gradient at the earth's surface is 23·6° K./km.

According to Emden the small temperature gradients above 10 km. are to be connected with the nearly isothermal stratosphere, and $\theta(\infty)$ should be the stratosphere temperature. However, this is not correct. It can be shown that the small gradients above 10 km. are a simple consequence of the assumed fact that the opacity of the atmosphere above this level is negligible, while fig. 53 shows that, in reality, the stratosphere as a whole still absorbs strongly in the major absorption bands. Further, according to Emden, the stratosphere temperature should vary as the fourth root of the incident solar energy, and therefore it should be greatest in the tropics and smallest in the Arctic, whereas the reverse is observed in practice. The model used here is in fact far too crude to deal with the properties of the atmosphere near to the tropopause, but the results for the surface layers are not altered qualitatively by more precise treatments.

The notable feature of the surface layers is their extreme instability, with a starting lapse rate nearly $2\frac{1}{2}$ times the dry adiabatic and an unstable discontinuity of $19°$ K. Thus violent convection will begin at the surface and, according to the figures used here, will spread upwards at least 1 or 2 km.

The question now arises as to how high the convective layer, which we will identify with the troposphere, can extend. It will obviously extend to that height which causes the starting temperature gradient in the lower stratosphere to be just stable. Now Milne (1922) has shown that, if the troposphere temperatures are given, then in general there will be a temperature discontinuity between troposphere and stratosphere similar to that which exists at the ground in the above example. In particular, Goody (1949) showed that, if the starting temperature gradient in the lower stratosphere is just stable, then the discontinuity is negative and unstable. This will cause the surface of division to rise still farther until a height is reached where there is no discontinuity of temperature; it can be shown that this is the condition for stable equilibrium, and that it allows only certain definite values for the tropopause height. Moreover, it can be shown that, with this stable equilibrium between troposphere and stratosphere, there must be a discontinuity of temperature gradient at the tropopause with the more positive gradient in the stratosphere.

This general picture so closely resembles a description of a normal tropopause that it is very hard to doubt that herein lies the essential reason for the observed temperature structure in the region of the tropopause, although, since no allowance has been made for winds and horizontal heat transfer, it cannot be expected to be very exact. Using the best data available at the time, Goody (1949) calculated the temperature at the tropopause upon the following simple model. Direct absorption of solar radiation near the tropopause is neglected and the earth's surface temperature and the temperature gradient in the troposphere are assumed to be known; the tropopause height required to avoid a temperature discontinuity at the tropopause can then be calculated approximately. Carbon dioxide, water vapour and ozone were all found to be of importance, and the most interesting feature of the calculation was that the stratosphere temperatures increased as ground temperatures decreased. It is difficult to assign a simple reason for this effect, although it is mainly caused by the water-vapour concentration at the tropopause being strongly correlated with air temperature through the observed fact that the percentage humidity at the tropopause is fairly constant. The relation found between ground temperature and tropopause temperature is very similar to the relation between these two temperatures observed at different latitudes, which suggests that this may be the reason for the observed latitude variation of tropopause temperature.

The seasonal variation of tropopause temperature has also been calculated on this model, following a suggestion of Dobson and others (1946) that variations of the ozone concentration in the lower stratosphere are responsible for the change of phase of the annual temperature wave at the tropopause. It is fairly obvious that an extra variable of this nature, considered together with the seasonal change of troposphere temperature, can be responsible for unexpected phase relations at the tropopause. However, Goody (1950b) has shown that, although the seasonal variation of ozone can produce an effect of importance, the phase relations are incorrect if the direct absorption of solar radiation at the tropopause is neglected. The direct absorption of solar radiation can, by itself, lead to a qualitative account of the seasonal variation of tropopause temperature, provided that the contribution to atmospheric heating by the Hartley and Huggins bands exceeds that of the Chappuis bands. This is contrary to the results of Karandikar's calculations near to the tropopause, but the reader should be reminded at this stage that all existing calculations are far from exact, and it appears, from what has been said, that the explanation of this particular effect may depend upon rather precise solutions.

The reason for the discontinuous change of temperature gradient at the tropopause has already been discussed. There has, however, been some discussion as to why the gradient in the lower stratosphere can be positive. The simple solution given above for the radiative equilibrium of a 'grey' atmosphere can, as a direct result of the equation of transfer, only lead to negative temperature gradients under all circumstances. But it has been pointed out, qualitatively by Dobson and others (1946) and quantitatively by Goody (1949), that this difficulty disappears if more than one absorbing component takes part in the radiative equilibrium,† or if there is direct absorption of solar radiation in the lower stratosphere. These effects would presumably be apparent from Gowan's computations if he were to break them down into detail.

With this brief description of synthetic models of the atmosphere we conclude this last chapter and our survey of the physics of the stratosphere. A complete understanding of all stratospheric processes requires, as fundamentals, more precise data upon stratospheric composition than those described in Chapters III and IV, more precise temperature measurements than those given in Chapter II, and a detailed treatment of the integral equations of §6.13. It will then be possible to calculate the radiation budget of the stratosphere in detail, and hence to discover the origin of the atmospheric motions described in Chapter V. This formidable programme will not be accomplished without a great deal of effort by many future workers.

† King (1952) points out that, if the lower boundary to a region in radiative equilibrium is a black body, temperature gradients are invariably negative even if several absorbing constituents are present. The system considered here has, however, the non-black troposphere as its lower boundary, and this is an essential point of the simple treatment of Dobson and others, which although crude is unobjectionable.

176

REFERENCES

ADEL, A. (1939). *Astrophys. J.* **89**, 1.
ADEL, A. (1947a). *Phys. Rev.* **71**, 806.
ADEL, A. (1947b). *Astrophys. J.* **105**, 406.
ADEL, A. (1949). *Rep. Engng. Res. Inst., Univ. Michigan.*
ADEL, A. (1950). *Cent. Proc. R. Met. Soc.* 5.
ARCHENOLD, F. S. (1928). *Weltall*, Jg. 27.
BABCOCK, H. D. (1923). *Astrophys. J.* **57**, 209.
BANNON, J. K. (1951). *Met. Mag., Lond.*, **80**, 262.
BARBIER, D. and CHALONGE, D. (1939). *J. Phys. Radium*, Ser. 7, **10**, 113.
BARBIER, D., CHALONGE, D. and VASSY, E. (1935). *Rev. Opt.* **14**, 425.
BARBIER, D., CHALONGE, D. and VIGROUX, E. (1942). *C.R. Acad. Sci., Paris*, **214**, 983.
BARRET, E. W., HERNDON, L. R. and HOWARD, J. C. (1950). *Tellus*, **2**, 302.
BARRET, E. W. and SUOMI, V. E. (1949). *J. Met.* **6**, 273.
BATES, D. R. and WITHERSPOON, A. E. (1952). *Mon. Not. R. Astr. Soc.* **112**, 101.
BECKER, G. E. and AUTLER, S. H. (1946). *Phys. Rev.* **70**, 300.
BIRGE, R. T. (1941). *Rep. Progr. Phys.* **8**, 90.
BJERKNES, V., BJERKNES, J., SOLBERG, H. and BERGERON, T. (1933). *Physikalische Hydrodynamik.* Berlin. Springer.
BOWEN, G. I. and REGENER, V. H. (1951). *J. Geophys. Res.* **56**, 307.
BRASEFIELD, C. J. (1948). *J. Met.* **6**, 273.
BRASEFIELD, C. J. (1950). *J. Met.* **7**, 66.
BREWER, A. W. (1949). *Quart. J. R. Met. Soc.* **75**, 351.
BROOKS, C. E. P., DURST, C. S., CARUTHERS, N., DEWAR, D. and SAWYER, J. S. (1950). *Geophys. Mem., Lond.*, no. 85.
BUDDEN, K. G., RADCLIFFE, J. A. and WILKES, M. V. (1939). *Proc. Roy. Soc. A*, **171**, 188.
CABANNES, J. and DUFAY, J. (1925). *C.R. Acad. Sci., Paris*, **181**, 302.
CARPENTER, T. M. (1939). *J. Amer. Chem. Soc.* **59**, 358.
CHACKETT, K. F., PANETH, F. A., REASBECK, F. and WIBORG, B. S. (1951). *Nature, Lond.*, **168**, 358.
CHAPMAN, S. (1930). *Mem. R. Met. Soc.* **3**, 103.
CHAPMAN, S. (1936). *Rep. Progr. Phys.* **3**, 42.
CHAPMAN, S. (1950). *Bull. Amer. Met. Soc.* **31**, 288.
CHAPMAN, S. (1951). *Proc. Phys. Soc. B*, **64**, 833.
CHAPMAN, S. and BARTELS, J. (1940). *Geomagnetism.* Oxford University Press.
COBLENZ, W. W. and STAIR, R. (1939). *Bur. Stand. J. Res., Wash.*, **22**, 573.
COBLENZ, W. W. and STAIR, R. (1941). *Bur. Stand. J. Res., Wash.*, **26**, 261.
COWLING, T. G. (1950). *Phil. Mag.* **41**, 109.
COX, E. F., ATANSOFF, J. V., SNAVELY, B. L., BEECHER, D. W. and BROWN, J. (1949). *J. Met.* **6**, 300.

CRAIG, R. A. (1950). *Met. Mon. Amer. Met. Soc.* **1**, no. 2.

CRARY, A. F. (1950). *J. Met.* **7**, 233.

CROSS, P. C., HAINER, R. M. and KING, G. W. (1944). *J. Chem. Phys.* **12**, 210.

CROSSLEY, A. F. (1934). *Met. Mag.* **69**, 30.

CURTIS, A. (1952). Private communication.

DEWAR, D. and SAWYER, J. S. (1947). M.R.P. 334, Air Ministry, London.

DIETRICHS, H. (1949). *Ber. dtsch. Wetterdienstes U.S. Zone*, 86.

DINES, L. G. H. (1936). *Quart. J. R. Met. Soc.* **62**, 379.

DOBSON, G. M. B. (1920). *Quart. J. R. Met. Soc.* **46**, 54.

DOBSON, G. M. B. (1930). *Proc. Roy. Soc.* A, **129**, 411.

DOBSON, G. M. B. (1931). *Proc. Phys. Soc.* **43**, 324.

DOBSON, G. M. B., BREWER, A. W. and CWILONG, B. M. (1946). *Proc. Roy. Soc.* A, **185**, 144.

DOBSON, G. M. B. and HARRISON, D. N. (1926). *Proc. Roy. Soc.* A, **110**, 660.

DOBSON, G. M. B., HARRISON, D. N. and LAWRENCE, J. (1927). *Proc. Roy. Soc.* A, **114**, 521.

DUFAY, J. (1936). *Quart. J. R. Met. Soc.* **62**, Suppl. 27.

DURAND, E. (1949). *The Atmospheres of the Earth and Planets.* Chicago University Press.

DÜTSCH, H. U. (1946). Dissertation, Univ. of Zurich.

EDDINGTON, A. S. (1929). *Mon. Not. R. Ast. Soc.* **89**, 623.

EHMERT, A. (1949a). *Ber. dtsch. Wetterdienstes U.S. Zone*, 11.

EHMERT, A. (1949b). *Z. Naturf.* **46**, 321.

ELSASSER, W. M. (1942). *Harvard Met. Studies*, no. 6.

ELTERMAN, L. (1951). *J. Geophys. Res.* **56**, 509.

EMDEN, R. (1913). *S.B. Akad. Wiss. München.*

EXPLORER II (1938). *Nature, Lond.*, **141**, 270.

FABRY, C. (1950). *L'Ozone Atmospherique.* Ed. du Centre Nat. de la Rech. Sci. Paris.

FABRY, C. and BUISSON, M. (1913). *J. Phys. Radium*, Ser. 5, **3**, 196.

FABRY, C. and BUISSON, M. (1921). *J. Phys. Radium*, Ser. 6, **2**, 197.

FLÖHN, H. and PENNDORF, R. (1950). *Bull. Amer. Met. Soc.* **31**, 71.

FOWLER, A. and STRUTT, HON. R. J. (1917). *Proc. Roy. Soc.* A, **93**, 557.

GASSIOT COMMITTEE (1942). *Rep. Progr. Phys.* **9**, 1.

GASSIOT COMMITTEE (1948). Physical Society.

GAUZIT, J. and GRANDMONTAGNE, R. (1942). *C.R. Acad. Sci., Paris*, **214**, 799.

GERSON, N. C. (1951). *Rep. Progr. Phys.* **14**, 719.

GLÜCKAUF, E. (1944). *Nature, Lond.*, **153**, 620.

GLÜCKAUF, E. (1945a). *Quart. J. R. Met. Soc.* **71**, 307.

GLÜCKAUF, E. (1945b). *Proc. Roy. Soc.* A, **185**, 98.

GLÜCKAUF, E. (1947). *Proc. Phys. Soc.* **59**, 344.

GLÜCKAUF, E. and PANETH, F. A. (1945). *Proc. Roy. Soc.* A, **185**, 89.

GLÜCKAUF, E., HEAL, H. G., MARTIN, G. R. and PANETH, F. A. (1944). *J. Chem. Soc. Trans.* 1.

GOLD, E. (1909). *Proc. Roy. Soc.* A, **82**, 43.

GOLDBERG, L. (1950). *Rep. Progr. Phys.* **13**, 24.

GOLDBERG, L. (1951). *Astrophys. J.* **113**, 567.

GOLDIE, A. H. R. (1923). *Quart. J. R. Met. Soc.* **49**, 6.

GOODY, R. M. (1949). *Proc. Roy. Soc.* A, **197**, 872.

GOODY, R. M. (1950a). *Cent. Proc. R. Met. Soc.* 9.

GOODY, R. M. (1950b). Dissertation, Cambridge University.

GOODY, R. M. (1952). *Quart. J. R. Met. Soc.* **78**, 165.

GOODY, R. M. and ROBINSON, G. D. (1951). *Quart. J .R. Met. Soc.* **75**, 161.

GOODY, R. M. and WORMELL, T. W. (1951). *Proc. Roy. Soc.* A, **209**, 178.

GÖTZ, F. P. W. (1931a). *Beitr. Geophys.* **31**, 119.

GÖTZ, F. P. W. (1931b). *Ergebn. kosm. Phys.* **1**, 205.

GÖTZ, F. P. W. (1938). *Ergebn. kosm. Phys.* **2**, 253.

GÖTZ, F. P. W. (1944). *Vjschr. naturf. Ges. Zürich*, **89**, 250.

GÖTZ, F. P. W., MEETHAM, A. R. and DOBSON, G. M. B. (1934). *Proc. Roy. Soc.* A, **145**, 416.

GOWAN, E. H. (1947). *Proc. Roy. Soc.* A, **190**, 219, 227.

GREENSTEIN, J. (1949). *The Atmospheres of the Earth and Planets.* Chicago University Press.

HAGELBARGER, D. W., LOH, L. T., WEILL, H. W., NICHOLS, M. H. and WENTZEL, E. A. (1951). *Phys. Rev.* **82**, 107.

HARANG, L. (1951). *The Aurorae.* London: Chapman and Hall.

HAVENS, R. J., KOLL, R.T. and LaGow, H.E. (1952). *J. Geophys. Res.* **57**, 59.

HEISENBERG, W. (1948). *Z. Phys.* **124**, 628.

VAN DER HELD, E. F. M. (1931). *Z. Phys.* **70**, 508.

HENRY, P. S. H. (1932). *Proc. Camb. Phil. Soc.* **28**, 249.

HERGESELL, H. (1919). *Arb. preuss. aero. Obs.* **13**.

HERMAN, R. (1945). *C.R. Acad. Sci., Paris*, **220**, 878.

HERZBERG, G. (1950). *Spectra of Diatomic Molecules.* New York: Van Nostrand.

HETTNER, G., POHLMANN, R. and SCHUHMACHER, H. J. (1935). *Z. Phys.* **91**, 372.

HOPF, E. (1934). *Mathematical Problems of Radiative Equilibrium.* Cambridge University Press.

HOUGHTON, H. C. (1951). *J. Met.* **8**, 270.

HOYLE, F. (1949). *Quart. J. R. Met. Soc.* **75**, 161.

VAN DE HULST, H. C. (1949). *The Atmospheres of the Earth and Planets.* Chicago University Press.

HUMPHREYS, W. J. (1933). *Mon. Weath. Rev., Wash.*, **61**, 228.

JEANS, J. H. (1921). *The Dynamical Theory of Gases.* Cambridge University Press.

JOHNSON, F. S., PURCELL, J. D., TOUSEY, R. and WATANABE, K. (1952). *J. Geophys. Res.* **57**, 157.

JUNGE, C. (1938). *Ann. Hydrogr., Berl.*, **3**, 104.

KAPLAN, L. D. (1950). *J. Chem. Phys.* **18**, 186.

KAPLAN, L. D. (1952). *J. Met.* **9**, 139.

KARANDIKAR, R, V. (1946). *Proc. Indian Acad. Sci.* **23**, 70.

KARANDIKAR, R. V. and RAMANATHAN, K. R. (1949). *Proc. Indian Acad. Sci.* **29**, 330.

KAY, R. H. (1951). *IXth Congr. U.G.G.I. Brussels.*

KELLOGG, W. W. and SCHILLING, G. (1950). Unpublished report to the U.S. Weather Bureau.

KELLOGG, W. W. and SCHILLING, G. (1951). *J. Met.* **8**, 222.

KING, J. I. (1952). *J. Met.* **9**, 311.

KOSCHMEIDER, H. (1936). *Dtsch. Met. Jb.* **5**, H. 13.

KUIPER, G. P. (1949). *The Atmospheres of the Earth and Planets.* Chicago University Press.

LETTAU, H. (1947). *Met. Rdsch.* **1**, 5.

LINDEMANN, F. A. and DOBSON, G. M. B. (1923). *Proc. Roy. Soc.* A, **102**, 411.

LINK, F. (1933). *Rep. Brit. Ass.* **9**, 227.

LINK, F. (1934). *J. obs.* **17**, 161.

MANNING, L. A., VILLARD, O. G. and PETERSEN, A. M. (1950). *Tech. Rep.* no. 22, Stamford University, California.

MAPLE, E., BOWEN, W. A. and SINGER, S. F. (1950). *J. Geophys. Res.* **55**, 115.

MARIS, H. B. (1928). *Terr. Magn. Atmos. Elect.* **33**, 223.

McQUEEN, J. H. (1950). *Phys. Rev.* **80**, 110.

MECKE, R. (1931). *Trans. Faraday Soc.* **27**, 375.

MEETHAM, A. R. (1936). *Quart. J. R. Met. Soc.* **62**, Suppl. p. 59.

MEINEL, A. B. (1951). *Rep. Progr. Phys.* **14**, 121.

MILNE, E. A. (1922). *Phil. Mag.* **144**, 872.

MILNE, E. A. (1930). *Handbuch der Astrophysik,* III (i), p. 162. Berlin: Springer.

MITRA, S. K. (1948). *The Upper Atmosphere.* The Royal Asiatic Society of Bengal.

MOELWYN-HUGHES, E. A. (1940). *Physical Chemistry.* Cambridge University Press.

MÖLLER, F. (1938). *Met. Z.* **5**, 161.

MURGATROYD, R. J. and CLEWS, C. H. B. (1949). *Geophys. Mem., Lond.,* no. 83.

NAMIAS, J. and CLAPP, P. F. (1949). *J. Met.* **6**, 330.

NEWELL, H. E. (1948). *Naval Res. Lab. Rep.* R 3294.

NEWELL, H. E. (1950). *Trans. Amer. Geophys. Un.* **31**, 15.

NEWELL, H. E. and SIRY, J. (1946). *Naval Res. Lab. Rep.* R 3030.

NICOLET, M. (1945). *Mém. Inst. mét. Belge,* **19**.

NORMAND, C. W. B. and KAY, R. H. (1952). *J. Sci. Instrum.* **29**, 33.

NY TSI-ZE and CHOONG SHIN PIAW (1932). *C.R. Acad. Sci., Paris,* **195**, 309.

NY TSI-ZE and CHOONG SHIN PIAW (1933). *C.R. Acad. Sci., Paris,* **196**, 916.

OLIVIER, C. P. (1947). *Proc. Amer. Phil. Soc.* **91**, no. 4, 315.

PAETZOLD, H. K. (1952). *J. Atmos. Terr. Phys.* **2**, 183.

PANETH, F. A. (1939). *Quart. J. R. Met. Soc.* **65**, 304.

PEDERSEN, E. F. (1942). *Met. Ann.* (Norwegian Met. Inst.), **1**, no. 6.

PEKERIS, C. L. (1937). *Proc. Roy. Soc.* A, **158**, 650.

PENNDORF, R. (1947). *F.I.A.T. Rev. Germ. Sci.* Meteorology and Physics of the Atmosphere, p. 222.

PHILLIPS, G. J. (1952). *J. Atmos. Terr. Phys.* **2**, 141.

PLASS, G. N. and WARNER, D. (1952). *J. Met.* **9**, 333.

PLYLER, E. K. and BARKER, E. F. (1931). *Phys. Rev.* **38**, 1827.

PRIESTLEY, C. H. B. (1944). *S.D.T.M.* no. 107, Air Ministry, London.

RAMANATHAN, K. R. and KARANDIKAR, R. V. (1949). *Quart. J. R. Met. Soc.* **75**, 257.

RANDALL, H. M., DENNISON, D. M., GINSBERG, N. and WEBER, L. R. (1937). *Phys. Rev.* **52**, 160.

REED, R. J. (1950). *J. Met.* **7**, 263.

REED, R. J. and JULIUS, A. L. (1951). *IXth Congr. U.G.G.I.* Brussels.

REGENER, E. (1935). *Beitr. Phys. frei. Atmos.* **22**, 247.

REGENER, E. (1939/40). *Schr. dtsch. Akad. Luftfahrt.* **46**, 5.

REGENER, E. (1949). *Ber. dtsch. Wetterdienstes U.S. Zone*, 11.

REGENER, E. and REGENER, V. H. (1934). *Phys. Z.* **35**, 778.

REGENER, V. H. (1938). *Z. Phys.* **109**, 642.

REGENER, V. H. (1951). *Nature, Lond.*, **167**, 276.

ROBERTS, O. F. T. (1930). *Proc. Roy. Soc. Edinb.* **50**, 225.

ROSSBY, C.-G. (1949). *The Atmospheres of the Earth and Planets.* Chicago University Press.

ROSSELAND, S. and STEENSHOLT, G. (1932). *Publ. Univ. Obs. Oslo*, **1**, 7.

SCHRÖDINGER, E. (1917). *Phys. Z.* **18**, 445.

SCHRÖER, E. (1949). *Ber. dtsch. Wetterdienstes U.S. Zone*, 11.

SCRASE, F. J. (1951). *Quart. J. R. Met. Soc.* **77**, 483.

SHELLARD, H. C. (1949). *Met. Mag.* **78**, 341.

SLOBAD, R. L. and KROGH, M. E. (1950). *J. Amer. Chem. Soc.* **72**, 1175.

SPITZER, L. (1949). *The Atmospheres of the Earth and Planets.* Chicago University Press.

STEWART, B. (1882). *Encyclopædia Britannica*, 9th ed., p. 36.

STÖRMER, C. (1933). *Univ. Obs. Oslo Publ.* **6**.

STÖRMER, C. (1948). *Weather*, **3**, 13.

STRONG, J. (1941). *J. Franklin Inst.* **231**, 121.

STRONG, J. (1949). Report to the O.N.R. Washington on Contract N5-ori-166 Task order V.

STRONG, J. and PLASS, G. N. (1950). *Astrophys. J.* **112**, 365.

STRUTT, HON. R. J. (1918). *Proc. Roy. Soc.* A, **94**, 260.

SWINGS, P. (1949). *The Atmospheres of the Earth and Planets.* Chicago University Press.

UNIVERSITY OF CHICAGO (1947). *Bull. Amer. Met. Soc.* **28**, 295.

VASSY, E. (1935). *C.R. Acad. Sci., Paris*, **202**, 1426.

VASSY, E. (1937). *Annales de Phys., Paris*, Ser. 11, **8**, 679.

VASSY, E. and VASSY, A. (1938). *C.R. Acad. Sci., Paris*, **207**, 1232.

VEGARD, L. (1937). *Phil. Mag.* **24**, 588.

VESTINE, E. H. (1934). *J. R. Astr. Soc. Can.* **28**, 249, 303.

VAN VLECK, J. H. and WEISSKOPF, V. F. (1945). *Rev. Mod. Phys.* **17**, 227.

WARFIELD, C. N. (1947). *Tech. Notes Nat. Adv. Comm. Aero., Wash.*, no. 1200.

WATANABE, K. (1943). *J. Franklin Inst.* **236**, 461.

WEXLER, H. (1950). *Tellus*, **2**, 263.

WHIPPLE, F. J. W. (1923). *Nature, Lond.*, **111**, 187.

WHIPPLE, F. J. W. (1935). *Quart. J. R. Met. Soc.* **61**, 285.

WHIPPLE, F. L. (1943). *Rev. Mod. Phys.* **15**, 246.
WHIPPLE, F. L. (1952). *Bull. Amer. Met. Soc.* **33**, 13.
WHIPPLE, F. L., JACCHIA, L. and KOPAL, Z. (1949). *The Atmospheres of the Earth and Planets.* Chicago University Press.
WILKES, M. V. (1949). *Oscillations of the Earth's Atmosphere.* Cambridge University Press.
WILSON, M. K. and BADGER, R. M. (1948). *J. Chem. Phys.* **16**, 741.
WULF, O. R. (1932). *Phys. Rev.* **41**, 375.
WULF, O. R. and DEEMING, L. S. (1936). *Terr. Magn. Atmos. Elect.* **41**, 229, 375.
YAMAMOTO, G. and ONISHI, G. (1949). *Sci. Rep. Tôhoku Univ.*, Geophys., **1**, 5.
YARNELL, J. and GOODY, R. M. (1952). *J. Sci. Instrum.* **29**, 352.
ZENER, C. (1931 *a*). *Phys. Rev.* **37**, 556.
ZENER, C. (1931 *b*). *Phys. Rev.* **38**, 277.

The following references, together with an indication of their subject-matter, may prove useful to a reader who wishes to know more about some of the problems which have been mentioned.

CHAPMAN (1951). A brief survey of recent work on the upper atmosphere.
CHAPMAN and BARTELS (1940). The relationships between atmospheric and magnetic phenomena.
CRAIG (1950). A review of the ozone problem.
DOBSON and others. (1946). Several contributions to the meteorology of the stratosphere.
FABRY (1950). A book devoted to the ozone problem.
GASSIOT COMMITTEE (1942). A number of articles about the physics and chemistry of the atmosphere.
GASSIOT COMMITTEE (1948). Report of an international conference on nightglow and auroral emissions.
GERSON (1951). A critical review of the determination of ionospheric temperatures.
GOODY and ROBINSON (1951). A review of atmospheric radiation problems.
HARANG (1951). A book dealing with all aspects of the aurorae.
KUIPER (1949). A collection of review articles on many atmospheric problems, including articles by Durand (rocket research), Swings (nightglow and auroral emission), Spitzer (the atmosphere above 300 km.), Greenstein (rocket research), van de Hulst (scattering), Whipple and others (meteors).
MEINEL (1951). A comprehensive review of nightglow and auroral spectra.
MITRA (1948). An excellent book upon all problems of the upper atmosphere.
NEWELL (1950). A review of rocket research.
WHIPPLE (1952). A summary of American upper atmosphere research, 1949–52.
WILKES (1949). A monograph about atmospheric oscillations.

INDEX

Abnormal propagation of sound, 30–6

Absorption bands of gases,
  in the infra-red spectrum, 66–8, 77–80, 84–6, 107–12, 145, 151–75
  in the ultra-violet spectrum, *see* Hartley bands, Huggins bands, Nitrogen, Oxygen, Ozone
  in the visible spectrum, 50–3; *see also* Chappuis bands

Accretion theory, 142

Adel, 47, 79, 108, 144, 156

Adiabatic compression errors, 14, 41

Adiabatic lapse rate, 1, 125, 141, 173

Aerobee rocket, 14–18, 75

Aircraft used for atmospheric research, 13, 61, 66–7, 139

Air masses, transfer of, 26, 106–7, 116–17

Anticyclones, 2, 116

Archenold, 133

Argon, 73

Aurorae, 4, 8, 48–53, 55

Babcock, 50

Balloons used for stratosphere research, 1, 9–13, 27, 57, 62–5, 70–4, 139

Band intensity, 147, 154

Bannon, 139

Barbier and Chalonge. 106

Barbier, Chalonge and Vassy, 103, 106

Barbier, Chalonge and Vigroux, 100

Barret, Herndon and Howard, 63

Barret and Suomi, 30

Bates and Witherspoon, 56, 70, 80

Becker and Autler, 156

Beer's law, 84, 164

Bellows gauges, 43

Birge, 20

Bjerknes, Bjerknes, Solberg and Bergeron, 126

Boltzmann's law, 146

Bowen and Regener, 102

Brasefield, 27–9, 130

Brewer, 76, 140

Brooks, Durst, Caruthers, Dewar and Sawyer, 127, 138

Bruce gun experiments, 129

Budden, Radcliffe and Wilkes, 49

Bumpiness, 139

Cabannes and Dufay, 81

Carbon dioxide,
  absorption and emission of radiation by, 7, 145–174
  concentration of, 10, 55, 70–2

Carbon monoxide, 56

Carpenter, 71

Catalytic destruction of ozone, 123

Chackett, Paneth, Reasbeck and Wiborg, 74

Chapman, 6, 20, 82, 117

Chappuis bands, 84–5, 100, 118, 144, 166, 174

Chemical analysis of ozone, 92, 101–3

Chicago school of meteorology, 126

Clausius-Clapeyron equation, 60

Clouds,
  mother-of-pearl, 68–9
  noctilucent, 46, 69–70, 130, 133–4, 138–9
  orographic, 69
  reflectivity of, 53, 148
  See also Water clouds *and* Ice clouds

Coblenz and Stair, 96

Collisions,
  electronic, 48
  molecular, 146–7

Composition,
  of the stratosphere, 55–125
  of the troposphere, 56

Convection,
  cellular, 139
  in the troposphere, 170–3

Cornu, 81

Corona, droplet, 68

Cosmic radiation, 10, 14

Cowling, 147, 154, 161–2

Cox, Atansoff, Snavely, Beecher and Brown, 31–4

Craig, 82, 84, 112, 118–24

Crary, 34–6, 128, 130, 132

Cross, Hainer and King, 152

Crossley, 1, 125

Curtis, 162–3

Density of air, 36–41, 115

Depressions, 2, 116

Dewar and Sawyer, 23–4

Dietrichs, 69

Diffuse radiation, 158; *see also* Scattering

Diffusion level, 70, 75–7

184 INDEX

Diffusive equilibrium, *see* Gravitational equilibrium
Dines, 71, 73
Dobson, 82–125
Dobson, Brewer and Cwilong, 25, 59–62, 72, 77, 170–5
Dobson and Harrison, 82, 86, 115
Dobson, Harrison and Lawrence, 115
Doppler line shape, 49–53, 154–61
Dufay, 104
Durand, 41, 97, 143
Dust, volcanic, 69–70
Dütsch, 118
Dynamo theory, 131, 136

E-region, 6, 49, 135
Eclipse of the moon, 99
Eddington, 148
Ehmert, 101, 104
Electrolytic indicator, 101, 102
Electron collisions, 21, 48–9
Electron density, 4, 5, 48–9
Elsasser, 158, 163
Elterman, 54
Emden, 170–3.
Emission temperature, 107
Excitation mechanisms, 52
Exosphere, 6
Explorer II, 9, 70, 72, 95
Extra-terrestrial constant, 87–9, 105

Fabry and Buisson, 82, 86–7
Flöhn and Penndorf, 6
Föhn wind, 103
Fowler and Strutt, 81
Fraunhofer lines, 95, 143
Frontal surfaces, 116
Frost-point hygrometer, 58–65
Frost-point temperature, 60–5

Gauzit and Grandmontagne, 53
Geostrophic wind, 136
Gerson, 48–9
Glaisher and Coxwell, 2
Glückauf, 58, 62, 70, 72
Glückauf and Paneth, 72, 73, 76
Glückauf, Heal, Martin and Paneth, 72, 73, 76
Gold, 170
Goldberg, 79–80
Goldie, 26
Goody, 66, 161, 170, 173–5
Goody and Robinson, 145
Goody and Wormell, 78, 147, 154, 156
Götz, 87, 90–3, 100

Götz, Meetham and Dobson, 82, 93, 168
Gowan, 167–5
Gradient wind, 126, 129, 135
Gravitational equilibrium, 76, 166
Gravitational separation, 31, 55, 58, 70, 73–7
Great Siberian meteor, 47
Greenstein, 14, 18
Grey absorption, 170, 175
Ground temperature, 171–2

Hagelbarger, Loh, Weill, Nichols and Wentzel, 57
Harang, 49
Hartley bands, 81–3, 110, 144, 166–7, 174
Havens, Koll and LaGow, 45
Hawke, 4
Heisenberg, 137
van der Held, 157–9
Helium, 10, 72–7, 140
Henry. 147
Hergesell, 170
Herman, 52
Herzberg, 51
    bands, 118
Hettner, Pohlmann and Schuhmacher, 85
Homopause, 6
Hopf, 171
Hopfield, 118
Houghton, 142
Hoyle, 142
Huggins bands, 47, 81–3, 96, 100, 103–4, 107, 110, 118, 144, 166, 174
van de Hulst, 100
Humidity, *see* Water vapour
Humphreys, 46–7
Hydrogen, 55, 56, 72
Hydrostatic equation, 21, 41
Hydroxyl radical, 55
Hygrometers,
    frost-point, manual, 59–62; automatic, 62–5
    gold-beater's skin, 58, 62
    wet-and-dry bulb thermometer, 58

Ice clouds, 4, 69
Inversion effect, 82, 90–5, 106, 120–1
Ionosphere, 6–7, 10, 14
    composition of, 49, 55, 76–7
    storms in, 114
    temperature of, 49–53
    winds in, 131, 135–136
Isothermal layer, 2, 3, 6

Jeans, 20–1
Jesse, 133, 139
Jet stream, 136–7
Johnson noise, 66
Johnson, Purcell, Tousey and Watanabe, 119
Junge, 139

Kaplan, 147, 158
Kaplan-Meinel band, 51
Karandikar, 166, 168, 174
Karandikar and Ramanathan, 90, 95
Kay, 104
Kellogg and Schilling, 44, 135, 141
King, 175
Kirchhoff's law, 107, 145–8, 171
Koschmeider, 139
Krakatoa explosion, 47

Lambert's law, 84–6, 171
Lettau, 140
Lifetimes of excited states, 146–7, 154–5
Lindemann and Dobson, 36–7
Line shapes, 154–7; see also Doppler line shape and Lorentz line shape
Line structure of molecular bands, 85–6
Line widths, 155–7
Linear law of absorption, 159
Link, 53, 101
Lorentz line shape, 79–80, 154–64, 168
Lunar spectra, 99–100, 108

McQueen, 57
Magnetic field of the earth, 7, 14, 114, 131
Manning, Villard and Petersen, 133
Maple, Bowen and Singer, 15
Maris, 76
Maxwellian distribution of velocities, 146, 156–7
Mean free path, 8, 21, 32, 76
Mecke, 117
Meetham, 115
Meinel, 49
Meinel bands, 51
Mesopause, 6
Meteor, 36–41
    trains, 130–3, 138
Methane, 77–80
Milne, 146, 170, 173
Mitra, 32, 76
Models,
    of absorption bands, 157–64
    of the atmosphere, 169–75

Moelwyn-Hughes, 147
Molecular bands, see Absorption bands
Molecular diameters, 21, 155
Molecular mass of air, 6, 8, 19, 50, 75
Molecular velocities, 21
Möller, 25
Monsoon winds, 128–32
Murgatroyd and Clews, 129

N.A.C.A. standard atmosphere, 39
Namias and Clapp, 128
Napier Shaw, 4
Neon, 72–7
Neutropause, 6
Newell, 14, 41–4
Newell and Siry, 16, 42
Nicolet, 123
Nightglow, 4, 8, 49–53, 55
Nitrogen, 4, 57, 73
    absorption of radiation by, 144
    atoms, 55
    bands in auroral spectra, 51
    iostopes, 57
Nitrous oxide, 77–9
    absorption bands, 77–9, 152, 164
Noctilucent clouds, see Clouds
Nomenclature, 3, 4, 6
Normand and Kay, 88–9
N.R.L. bead spectrometer, 97
Numerical computation of radiational heating, 150–1
Ny and Choong, 82, 87

Occlusions, 116
Olivier, 132
Oscillations of the atmosphere, 8, 47–8, 131, 136
Oxygen, 4, 57, 72, 117–20
    absorption bands of, 118, 120, 144, 163
    bands in auroral spectra, 51
    de-excitation of molecules of, 147
    deficits in the stratosphere, 58, 140
    photochemistry of, 46, 55, 117–20
Ozone, 4, 7, 10, 11, 81–124
    absorption spectrum of, 82–6, 144, 146, 152
    astronomical measurements of, 99–101
    centre of gravity of, 110–12
    chemical determination of, 101–3
    decay of, 122–3
    diurnal variation of, 115
    ionospheric storms and, 115

Ozone (*cont.*)
latitude and seasonal variation of amount of, 112–15, 121–4
magnetic phenomena and, 114
mean height of, 110, 112
photochemistry of, 55, 118–20
solar constant and, 114
stratosphere temperatures and, 120, 150, 164–75
sunspot number and, 114
temperature of layer of, 103–10
theoretical treatment of problem of, 82, 96, 99, 117–24
total amount of, 86–9
tropospheric, 103, 120–4
vertical distribution of, 90–103, 166, 168
weather and, 115–17

P, Q and *R* branches of molecular bands, 151–4
Paetzold, 100
Paneth, 58
Pedersen, 160–4
Pekeris, 47, 136
Penndorf, 140
Phillips, 135
Phillips gauge, 43
Photochemical theory, *see* Oxygen *and* Ozone
Piccard, 2
Pirani gauge, 43
for He and Ne measurements, 73
Planck's function, 107
Plass and Warner, 155
Plyler and Barker, 152
Potential temperature, 115
Pressure, 21
gauges, 1, 43
measurement of, 2, 12, 14, 41–6
scaling approximation, 162–3
static, 17, 41–2
ram, 41–2
Priestley, 26

Radiation, 142–7.
balance, 7, 47, 61, 66, 141
diagrams, 150–164
errors, 2, 13, 29, 30
shields, 29
transfer, 7, 145, 148
Radiative equilibrium, 167–75
Radio sondes, 10, 12, 34–5, 43–5, 96
Radio techniques, 5, 48
Radio waves, reflexion of, 7, 48–9
Radon, 56

Ramanathan and Karandikar, 87
Randall, Dennison, Ginsberg and Weber, 153
Rare gases, 49, 73–7
Rate coefficients, 118–20
Rayleigh scattering, *see* Scattering of light
Recovery of data, 10, 16, 17
Reed, 124
Reed and Julius, 123
Refraction of solar radiation, 100
Regener, E., 10–12, 62, 105
Regener, E. and V. H., 82, 95
Regener, V. H., 95
Residual rays, 110
Roberts, 158
Rockets, 14–18, 43, 97; *see also* Aerobee rocket, V 2 rocket *and* Viking rocket
air samples from, 57, 74
spectra from, 97, 143
temperature measurement from, 41–6
winds measured from, 131–2
Rossby, 127
Rosseland and Steensholt, 52

Scale height, 20, 48–9
Scattering of light, 148; *see also* Inversion effect
and atmospheric temperature, 53–7
from dust and water drops, 53, 56, 86–7, 89, 144
in spectroscopes, 89, 105, 108, 110
molecular, 53–4, 86, 90–95, 100, 105, 144
multiple, 93
plates, 95, 97
white, 87, 144
Schmidt camera, 37
Schrödinger, 31–3
Schröer, 118
Schumann bands, 118
Scrase, 27–9, 130
Searchlights, scattering from, 54
Shellard, 61
Shielding effect of the atmosphere, 172
Shock waves on rockets, 42
Sirius, spectrum of, 81
Slobad and Krogh, 79–80
Smithsonian Institute, 142
Sodium atoms, 55
Solar colour temperature, 143–4, 169
Solar constant, 114, 142–3
Solar control over atmospheric temperatures, 28, 29, 164–175

Solar flares, 144
Solar radiation, 117, 120, 142–75
Solid matter suspended in the atmosphere, 56
Sound velocity, 19, 30–6, 41, 147
Sound waves, refraction and absorption of, 30–6
Specific heats of air, 20–30
Spectrophotometer, 88
Spectroscopy, 10, 11, 14, 47, 56, 66, 77, 106, 108–12, 143
Spitzer, 49, 53, 76, 146
Square-root law of absorption, 159
Stability,
    hydrostatic, 3, 170–3
    of balloons, 11
    of rockets, 14, 17
Stefan's law, 171
Stewart, 136
Störmer, 68, 133–4, 139
Strong, 66, 84, 110
Strong and Plass, 160
Strutt, 81
Swings, 49

Teisserenc de Bort, 1–3
Telemetering, 14–17
Temperature,
    discontinuities, 171–4
    in the lower stratosphere, 22–9
    kinetic, 52–146
    latitude variation of, 25, 29, 40
    of the ozone layer, 103–6, 107–10
    seasonal variation of, 25–7, 28, 37–40
    short-period fluctuations of, 23–5, 107, 108, 116
    vibrational and rotational, 50–3
    See also Aurorae, Clouds, Collisions, Meteor, Nightglow, Oscillations of the atmosphere, Scale height, Scattering of light, Sound velocity and Rockets
Terrestrial radiation. 144, 145, 148, 167, 171–5
Thermometers, equilibrium with the atmosphere, 19, 29, 41
Three-body collisions, 117–18
Transitions,
    permitted and forbidden, 50
    spontaneous and induced, 146–7
Transmission functions, 158, 164
Turbulence, 136–41
    anisotropy of, 137, 139–40

diffusion by, 137
energy of, 137–8
spectrum of, 137
transfer by, 56, 125–41

University of Chicago, 126
Upper mixing layer, 6

V2 rocket, 14–18, 41–5, 132
Vassy, E., 84–5
Vassy, E. and Vassy, A., 106
Vegard, 50–1
Vegard-Kaplan bands, 51
Vestine, 69, 133–5, 138
Viking rocket, 14–18
van Vleck and Weisskopf, 155

Warm layer, 6
Warfield, 39
Watanabe, 88, 110–12
Water clouds, 4, 46
Water vapour, 1, 4, 7, 21, 58–70, 150, 168, 170, 174
    absorption bands, 66–8, 144–6, 152, 163, 167
    photo-chemical destruction of, 46, 55–6
    See also Hygrometers
Weather systems, 112
Wexler, 25
Whipple, F. J. W., 31–40, 129
Whipple, F. L., 37–9, 45
Whipple, F. L., Jacchia and Kopal, 37–40
Wilkes, 8, 47–8
Wilson and Badger, 85
Winds, 125–41
    diurnal variation of, 132, 135
    irregular changes of, 131
    in the lower stratosphere, 127–30
    from sound propagation experiments, 30–6, 128, 130–6
    near the tropopause, 125–7, 173
    in the upper stratosphere, 130–6
Wulf, 117
Wulf and Deeming, 118

Xenon, 56

Yamamoto and Onishi, 152, 154
Yarnell and Goody, 66

Zener, 146–7
Zones of audibility and silence, 31

Printed in the United States
By Bookmasters